Heiko Braak and Kelly Del Tredici

Neuroanatomy and Pathology of Sporadic Parkinson's Disease

With 29 Figures

Springer

Heiko Braak
Institute for Clinical Neuroanatomy
Goethe University Frankfurt
Theodor Stern Kai 7
60590 Frankfurt am Main
Germany
e-mail: braak@em.uni-frankfurt.de

Kelly Del Tredici
Institute for Clinical Neuroanatomy
Goethe University Frankfurt
Theodor Stern Kai 7
60590 Frankfurt am Main
Germany
e-mail: tredici@em.uni-frankfurt.de

ISSN 0301-5556
ISBN 978-3-540-79849-1 e-ISBN 978-3-540-79850-7

Library of Congress Control Number: 2008932111

9 8 7 6 5 4 3 2 1

springer.com

201
Advances in Anatomy, Embryology and Cell Biology

Reviews and critical articles covering the entire field of normal anatomy (cytology, histology, cyto- and histochemistry, electron microscopy, macroscopy, experimental morphology and embryology and comparative anatomy) are published in Advances in Anatomy, Embryology and Cell Biology. Papers dealing with anthropology and clinical morphology that aim to encourage cooperation between anatomy and related disciplines will also be accepted. Papers are normally commissioned. Original papers and communications may be submitted and will be considered for publication provided they meet the requirements of a review article and thus fit into the scope of "Advances". English language is preferred.

It is a fundamental condition that submitted manuscripts have not been and will not simultaneously be submitted or published elsewhere. With the acceptance of a manuscript for publication, the publisher acquires full and exclusive copyright for all languages and countries.

Twenty-five copies of each paper are supplied free of charge.

Manuscripts should be addressed to

Prof. Dr. F. **BECK**, Howard Florey Institute, University of Melbourne, Parkville, 3000 Melbourne, Victoria, Australia
e-mail: fb22@le.ac.uk

Prof. Dr. F. **CLASCÁ**, Department of Anatomy, Histology and Neurobiology,
Universidad Autónoma de Madrid, Ave. Arzobispo Morcillo s/n, 28029 Madrid, Spain
e-mail: francisco.clasca@uam.es

Prof. Dr. M. **FROTSCHER**, Institut für Anatomie und Zellbiologie, Abteilung für Neuroanatomie,
Albert-Ludwigs-Universität Freiburg, Albertstr. 17, 79001 Freiburg, Germany
e-mail: michael.frotscher@anat.uni-freiburg.de

Prof. Dr. D.E. **HAINES**, Ph.D., Department of Anatomy, The University of Mississippi Med. Ctr.,
2500 North State Street, Jackson, MS 39216–4505, USA
e-mail: dhaines@anatomy.umsmed.edu

Prof. Dr. N. **HIROKAWA**, Department of Cell Biology and Anatomy, University of Tokyo,
Hongo 7–3–1, 113-0033 Tokyo, Japan
e-mail: hirokawa@m.u-tokyo.ac.jp

Dr. Z. **KMIEC**, Department of Histology and Immunology, Medical University of Gdansk,
Debinki 1, 80-211 Gdansk, Poland
e-mail: zkmiec@amg.gda.pl

Prof. Dr. H.-W. **KORF**, Zentrum der Morphologie, Universität Frankfurt,
Theodor-Stern Kai 7, 60595 Frankfurt/Main, Germany
e-mail: korf@em.uni-frankfurt.de

Prof. Dr. E. **MARANI**, Department Biomedical Signal and Systems, University Twente,
P.O. Box 217, 7500 AE Enschede, The Netherlands
e-mail: e.marani@utwente.nl

Prof. Dr. R. **PUTZ**, Anatomische Anstalt der Universität München,
Lehrstuhl Anatomie I, Pettenkoferstr. 11, 80336 München, Germany
e-mail: reinhard.putz@med.uni-muenchen.de

Prof. Dr. Dr. h.c. Y. **SANO**, Department of Anatomy, Kyoto Prefectural University of Medicine,
Kawaramachi-Hirokoji, 602 Kyoto, Japan

Prof. Dr. Dr. h.c. T.H. **SCHIEBLER**, Anatomisches Institut der Universität,
Koellikerstraße 6, 97070 Würzburg, Germany

Prof. Dr. J.-P. **TIMMERMANS**, Department of Veterinary Sciences, University of Antwerpen,
Groenenborgerlaan 171, 2020 Antwerpen, Belgium
e-mail: jean-pierre.timmermans@ua.ac.be

List of Contents

1	**Prologue**	**1**
1.1	sPD and Parkinsonism	2
1.2	sPD Is a Proteinopathy Linked to the Development of Abnormal Intraneuronal Inclusions (Lewy Pathology: Lewy Neurites, Lewy bodies, Lewy plaques)	3
2	**Morphology of Lewy Pathology**	**8**
2.1	The Protein α-Synuclein	8
2.2	Protein Misfolding and Abnormal Aggregation of α-Synuclein	9
2.3	Light Microscopic Visualization of Lewy Pathology	10
2.3.1	Lewy Neurites	12
2.3.2	Granular Aggregations	13
2.3.3	Pale Bodies and Lewy Bodies	13
2.3.4	Lewy Plaques	14
2.4	Degradation of Lewy Pathology	14
2.5	Lewy Pathology Can Co-occur with Pathology Related to Other Disorders	15
3	**The Evolving Distribution Pattern of Lewy Pathology Associated with sPD Renders Neuropathological Staging Possible**	**15**
3.1	Incidental Lewy Pathology	15
3.2	Presymptomatic and Symptomatic Phases	15
3.3	Vulnerable Regions are Interconnected Anatomically	18
4	**Stage 1**	**20**
4.1	Pathology in Olfactory Structures	20
4.2	Basic Organization of the Medullary Autonomic Region	21
4.3	Involvement of Preganglionic Parasympathetic Projection Neurons	21
4.4	Pathology in the Enteric Nervous System	23
4.5	Is sPD Inducible by a Neurotropic Pathogen Akin to a Slow Virus?	23
4.6	Involvement of Preganglionic and Postganglionic Sympathetic Projection Neurons	29
4.7	Pathology in Lamina I of the Spinal Cord	29
5	**Stage 2**	**32**
5.1	Involvement of the Medulla Oblongata and Pontine Tegmentum	32
5.2	Basic Organization of the Gain Setting System	35
5.2.1	Lower Raphe Nuclei	35
5.2.2	Magnocellular Reticular Nuclei	35
5.2.3	Coeruleus–Subcoeruleus Complex	36
5.3	Potential Functional Consequences of the Lesions	37

6	**Stage 3**	**38**
6.1	Involvement of the Mesencephalic Tegmentum and Basal Forebrain	38
6.1.1	Central Subnucleus of the Amygdala	38
6.1.2	Substantia Nigra and Adjoining Nuclei	40
6.2	Basic Organization of Nonthalamic Nuclei with Diffuse Projections	43
6.2.1	Magnocellular Nuclei of the Basal Forebrain	44
6.2.2	Pedunculopontine Tegmental Nucleus	44
6.2.3	Tuberomamillary Nucleus	45
6.2.4	Upper Raphe Nuclei	45
6.3	Potential Functional Consequences of the Lesions	46
7	**Stage 4**	**48**
7.1	Involvement of the Amygdala and Thalamus	49
7.1.1	Amygdala	49
7.1.2	Thalamus	50
7.2	Basic Organization of the Cerebral Cortex and Limbic Circuit	52
7.2.1	Allocortex and Neocortex	52
7.2.2	Mesocortex	57
7.2.3	Limbic Circuit	58
7.3	Involvement of the Temporal Mesocortex and Hippocampal Formation	59
7.3.1	Temporal Mesocortex	59
7.3.2	Hippocampal Formation	60
7.4	Involvement of Insular, Subgenual, and Anterior Cingulate Areas	60
7.5	Potential Functional Consequences of the Lesions	61
8	**Stages 5 and 6**	**64**
8.1	Involvement of Neocortex and Basal Ganglia	65
8.1.1	Neocortex	65
8.1.2	Basal Ganglia	67
8.2	Basic Organization of the Basal Ganglia and Striatal Circuit	69
8.2.1	Striatum	69
8.2.2	Pallidum	71
8.2.3	Subthalamic Nucleus	72
8.2.4	Claustrum	73
8.3	Dot-Like Inclusions and Abnormally Altered Astrocytes in the Forebrain	73
8.4	Disruption of the Striatal Circuit	80
8.5	Potential Functional Consequences of Lesions	84
8.6	Pathologically Changed Components of the Cerebellar Circuit	85
9	**The Progression of the Cortical Lesions Mimics the Pattern of Myelination in Reverse Order**	**88**
10	**The Staging Hypothesis: Assumptions, Challenges, Potential**	**89**

Abstract

The proteinopathy sporadic Parkinson's disease (sPD) is the second most frequent degenerative disorder of the human nervous system after Alzheimer's disease. The α-synuclein inclusion body pathology (Lewy pathology) associated with sPD is distributed throughout the central, peripheral, and enteric nervous systems. The resulting nonrandom neuronal dysfunction and, in some regions, neuronal loss is reflected in a topographic distribution pattern of the Lewy pathology that, in the brain, can be staged. Except for olfactory structures and spinal cord constituents of the pain system, sensory components of the nervous system remain uninvolved or virtually intact. The most disease-related damage revolves around motor areas—particularly around superordinate centers of the limbic and visceromotor systems as well as portions of the somatomotor system. Vulnerable regions are interconnected anatomically and susceptible nerve cell types are not neurotransmitter-dependent. Not all clinical symptoms emerging in the course of sPD can be explained by a lack of dopamine in the nigrostriatal system. These include autonomic dysfunction, pain, hyposmia or anosmia, excessive daytime sleepiness, rapid eye movement (REM) sleep behavioral disorder, depression, anxiety, cognitive decline, and dementia. Against the background of the normal morphology and anatomy, the authors analyze the pathoanatomy of sPD in the nervous system at various neuropathological stages and summarize the potential functional consequences of the lesions.

1 Prologue

The synucleinopathy sporadic Parkinson's disease (sPD) is the second most frequent degenerative disorder of the human nervous system after Alzheimer's disease. The propensity for developing sPD exists in all ethnic groups worldwide, and the prevalence of the disorder increases considerably with age, thereby imposing an enormous social and economic burden on societies with increased life expectancy. The sPD-associated pathological process is progressive, does not go into remission, and can take decades to reach its culmination if it is not terminated prematurely by death owing to other causes (Lang and Lozano 1998a, b; Tanner and Aston 2000; de Lau and Breteler 2006; Hofman et al. 2006; Hirtz et al. 2007).

To date, the sPD-associated pathology is not known to affect the nervous systems of nonhuman vertebrates or to involve organ systems apart from the nervous system. The definition of sPD as a monosystemic disorder with preferential obliteration of dopaminergic neurons in the nigrostriatal system is inadequate because it is too narrow. The pathology is widely distributed throughout the entire nervous system, not only the central (CNS) but also the peripheral (PNS) and enteric nervous systems (ENS). Vulnerable nerve cell types cannot be distinguished from those that resist the pathological changes with respect to neurotransmitter type and functional systems. The topographical distribution pattern of the lesions in the brain evolves more or less symmetrically and is remarkably consistent across cases (Wakabayashi et al. 1988, 1990; Jellinger 1991, 2001; Fearnley and Lees 1994; Forno 1996; Lang and Lozano 1998a, b; Takahashi and Wakabayashi 2001, 2005; Jellinger and Mizuno 2003; Duda et al. 2007; Braak and Del Tredici 2008).

Sporadic PD is characterized by a variety of clinical signs, discrete at first and seemingly unrelated. For the most part, the symptoms begin insidiously and gradually worsen over time. Many emerge during the disease course that cannot be explained by a lack of dopamine in the nigrostriatal system. Among these are pain, hyposmia or anosmia, autonomic dysfunction, excessive daytime sleepiness, rapid eye movement (REM) sleep behavioral disorder, depression, anxiety, and cognitive decline (Calne et al. 1992; Rajput 1994; Dubois and Pillon 1997; Mesholam et al. 1998; Takatsu et al. 2000; Abbott et al. 2001, 2005, 2007; Benarroch 2006; Tissingh et al. 2001; Aarsland et al. 2003; Boeve et al. 2003, 2007; Fahn 2003; Hawkes 2003; Litvan et al. 2003, 2007; McDonald et al. 2003; Rascol et al. 2003; Emre 2004; Klockgether 2004; Koller and Tse 2004; Lang and Obeso 2004; Ponsen et al. 2004; Adler 2005; Adler and Thorpy 2005; Ahlskog 2005, 2007; Boeve and Saper 2006; Chaudhuri et al. 2006; Galvin et al. 2006; Haehner et al. 2007; Langston 2006; Ross et al. 2006; Idiaquez et al. 2007; Cersósimo and Benarroch 2008).

1.1 sPD and Parkinsonism

In the currently clinically recognizable phase of sPD, most patients present with signs of motor dysfunction, such as bradykinesia or hypokinesia, cogwheel rigidity, resting tremor, and postural instability (Parkinson 1817; Fahn 2003). The same clinical phenotype as sPD, however, can also appear as parkinsonism in other disorders associated with a sufficient reduction of dopamine in the nigrostriatal system. Familial forms of parkinsonism have been described (Nussbaum and Polymeropoulos 1997; Jellinger and Mizuno 2003; Giasson et al. 2006; Riess et al. 2006; Alegre-Abarrategui et al. 2007), and the syndrome may result following infection (Nelson and Paulson 2002; Cassarino et al. 2003), intoxication (Forno et al. 1993; Petrovitch et al. 2002; Sherer et al. 2003; Kamel et al. 2007), or head trauma (Bower et al. 2003; Goldman et al. 2006). In addition, parkinsonism may develop during metabolic disease (Taly et al. 2007), normal pressure hydrocephalus, or as a result of vascular-related changes (Stolze et al. 2001; Hughes et al. 2002; Sibon and Tison 2004). Finally, parkinsonism occurs in other neurodegenerative diseases, both in tauopathies and synucleinopathies (Table 1; Hughes et al. 1992; Tison 1997; Gelb et al. 1999; Giasson et al. 2000a; Dickson 2001; Dickson et al. 2007; Galvin et al. 2001; Litvan et al. 2003; Duyckaerts 2005; McKeith et al. 2005; Chade et al. 2006; Gilman 2006; Attems et al. 2007).

The existence of disease entities that elicit symptoms mimicking sPD makes differential diagnosis difficult, and autopsy-based assessment of sPD is still required to confirm not only the clinical diagnosis but also potentially presymptomatic individuals in the broader population at large (Litvan et al. 2007). Individuals who

Table 1 Synucleinopathies and tauopathies, diseases with Lewy pathology

Synucleinopathies	Pure autonomic failure (Hague et al. 1997; Arai et al. 2000; Kaufmann et al. 2001)
	Lewy body dysphagia (Kovári et al. 2007)
	Dementia with Lewy bodies (Kosaka et al. 1984; Ince and McKeith 2003; Geser et al. 2005)
	Sporadic Parkinson's disease (Jellinger and Mizuno 2003)
	Neurodegeneration with brain iron I (NBI1) (Arawaka et al. 1998; Galvin et al. 2000)
	Multiple system atrophy (Spillantini et al. 1998; Lantos and Quinn 2003; Galvin et al. 2001; Wenning et al. 2004)
	Amyotrophic lateral sclerosis (Kato et al. 2003)
Tauopathies	Progressive supranuclear palsy (Hauw et al. 1998; Hauw and Agid 2003; Dickson et al. 2007)
	Corticobasal degeneration (Revesz and Daniel 1998; Dickson and Litvan 2003)
	Down's syndrome (Lippa et al. 1999)
	Alzheimer's disease (Hansen 1997; Hamilton 2000; Attems et al. 2007)

display parkinsonism in the absence of the characteristic sPD-related pathological changes probably have one of the heterogeneous disorders listed in Table 1.

1.2 sPD Is a Proteinopathy Linked to the Development of Abnormal Intraneuronal Inclusions (Lewy Pathology: Lewy Neurites, Lewy bodies, Lewy Plaques)

The pathological process that underlies sPD is marked by the ongoing formation of proteinaceous α-synuclein immunoreactive inclusions (Lewy pathology) in especially vulnerable neuronal types. Of the many types of nerve cells within the human PNS/ENS/CNS, only a select few develop the inclusions while other types, directly in the vicinity of the affected ones, maintain their morphological and functional integrity. The resulting nonrandom neuronal dysfunction and, in some regions, neuronal loss is reflected by a distinctive topographic distribution pattern of the Lewy pathology. Sensory components of the human nervous system remain uninvolved or, for the most part, intact, with the exception of olfactory structures and portions of the pain system. The most disease-related damage revolves around motor

Table 2 Overview of the anatomical systems and regions involved in sporadic Parkinson's disease

Sensory centers	Olfactory: early and severe involvement
	Nociceptive: early involvement
	Somatosensory: mostly intact
	Viscerosensory: mostly intact
	Auditory and visual: uninvolved
Motor centers	Visceromotor: early and severe involvement
	Somatomotor: partial involvement
	Limbic: severe involvement
Stage 1	Anterior olfactory nucleus
	Olfactory bulb, olfactory tract
	Dorsal motor nucleus of the vagal nerve
	Intermediate reticular zone
Stage 2	(Plexuses of Meissner and Auerbach)
	Lower raphe nuclei
	Magnocellular reticular nuclei
	Coeruleus-subcoeruleus complex
Stage 3	(Spinal cord lamina I)
	Central subnucleus of the amygdala
	Olfactory tubercle, piriform cortex (olfactory system)
	Periamygdalear cortex (olfactory system)
	Medial entorhinal region (olfactory system)
	Substantia nigra, pars compacta
	Paranigral nucleus

(continued)

Table 2 (continued)

	Edinger Westphal nucleus
	Pigmented parabrachial nucleus
	Upper raphe nuclei
	Tuberomammillary nucleus
	Magnocellular basal forebrain nuclei
	Pedunculopontine tegmental nucleus
Stage 4	Interstitial nucleus of the terminal stria
	Cortical and basolateral amygdala
	Thalamic intralaminar nuclei
	Thalamic midline nuclei
	Anteromedial temporal mesocortex
	Ammon's horn, second sector (CA2)
	Insular and subgenual cortex
	Anterior cingulate cortex
	Ventral claustrum
Stage 5	High order sensory association neocortex
	Prefrontal neocortex
	Entorhinal region, CA1 and CA3 sectors
Stage 6	First order sensory association neocortex
	Premotor neocortex
	Primary sensory areas
	Primary motor field

areas—particularly around superordinate centers of the limbic and visceromotor systems as well as portions of the somatomotor system (Table 2; Jellinger 1991; Jellinger and Mizuno 2003; Braak et al. 1998, 2004; Braak and Del Tredici 2005, 2008; Bloch et al. 2006; Klos et al. 2006).

Fig. 1 **a** Lewy pathology develops in neuronal types that have some notable properties in common. Virtually all of them are projection neurons with an axon that is disproportionately long and slender in relation to the size of the parent cell body. **b–d** Selective neuronal vulnerability in sporadic Parkinson's disease. **b** Projection and local circuit neurons with short axons do not become involved. **c** Neurons with a thick-caliber myelin sheath are also resistant. A sturdy myelin sheath appears to have the following advantages: high conductivity, reduced energy expenditure, and less oxidative stress on the part of the parent neuron. The oligodendroglia cells that produce and sustain the myelin sheath stabilize the parent neuron by preventing pathological sprouting. **d** Projection neurons that have a long, thin-caliber axon that is poorly myelinated or unmyelinated are predisposed to Lewy body diseases. **e** Schematic representation of the natively unfolded presynaptic protein α-synuclein showing the lipid binding repeat region followed by a hydrophobic NAC (non-Aβ component of amyloid) region and a negatively charged carboxy terminal, which, when truncated, may be an important regulator of aggregation in vivo. α-Synuclein is most strongly expressed in the brain but also is expressed in skeletal muscle, the heart, lung, kidney, pancreas, and placenta. Dysfunction of the 140-amino acid protein plays a central role in Lewy body disorders

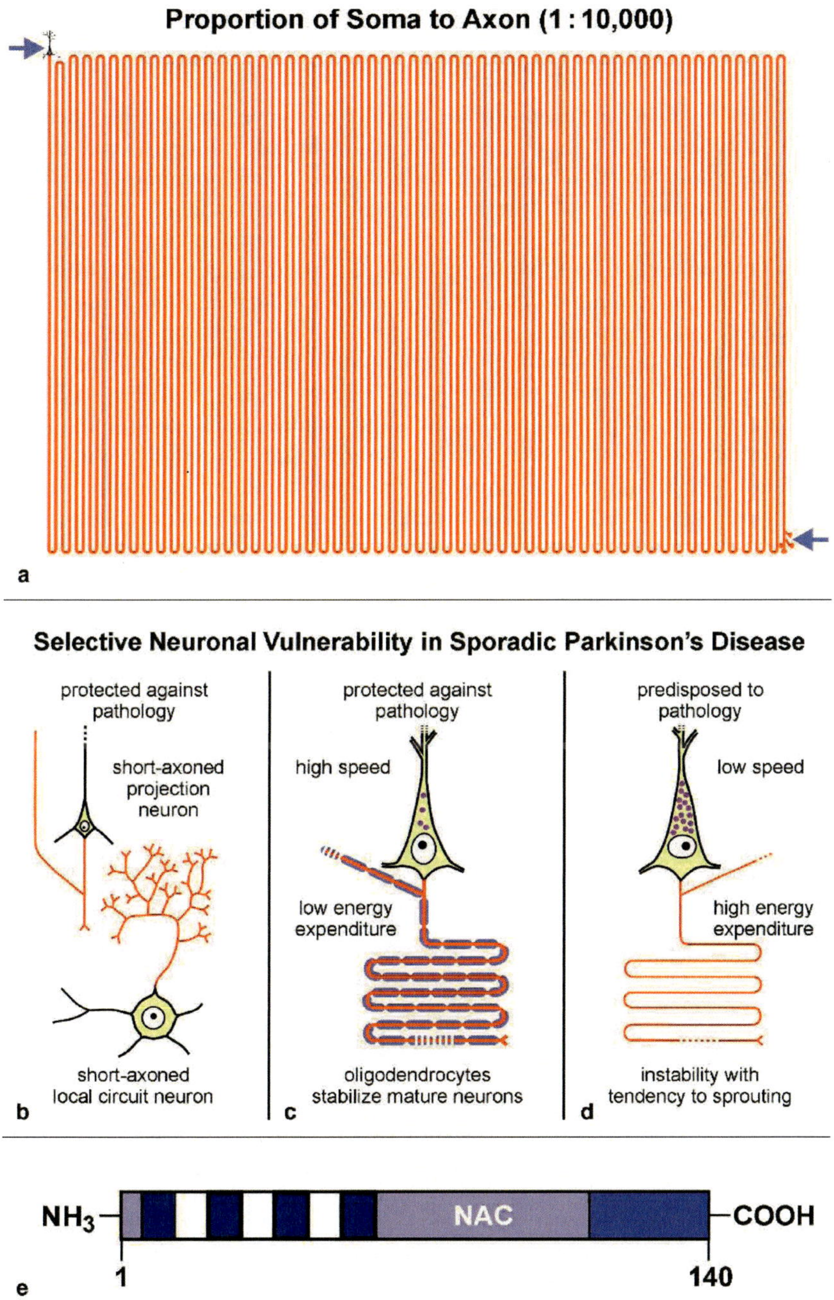
Proportion of Soma to Axon (1 : 10,000)
a
Selective Neuronal Vulnerability in Sporadic Parkinson's Disease
protected against pathology
short-axoned projection neuron
short-axoned local circuit neuron
b
protected against pathology
high speed
low energy expenditure
oligodendrocytes stabilize mature neurons
c
predisposed to pathology
low speed
high energy expenditure
instability with tendency to sprouting
d
NH3
NAC
COOH
1
140
e

The neuronal types prone to involvement have five properties in common: all contain the presynaptic protein α-synuclein and all belong to the class of projection neurons. Third, all contain either lipofuscin or neuromelanin granules (Gray and Woulfe 2005; Zecca et al. 2006). Fourth, only projection cells with a thin-caliber axon that is disproportionately long in relationship to the size of the soma demonstrate a pronounced tendency to develop the inclusion body pathology (Fig. 1a). Inasmuch as glutamatergic, GABAergic, dopaminergic, noradrenergic, serotonergic, histaminergic, and cholinergic projection cells become affected, the neurotransmitter type synthesized is not, in itself, an adequate criterion for assigning nerve cells to the ranks of the vulnerable neurons or for predicting which neurons are predisposed to sPD (Braak and Del Tredici 2008). With the exception of one recent report, Lewy pathology has not been found to date in short-axoned local circuit neurons (Fig. 1b; Mori et al. 2007). Nerve cells that lack α-synuclein, lipofuscin, or neuromelanin granules resist the formation of Lewy pathology and, similarly, nearly all projection cells with a short axon remain uninvolved, such as the small pyramidal cells of neocortical layers II and IV, the granule cells of the dentate fascia in the hippocampal formation, and the neurons of the presubicular parvocellular layer. Thus, the presence of lipofuscin or neuromelanin granules, as well as a disproportionately long and thin-caliber axon, appears to be among the prerequisites for the formation of sPD-related inclusion bodies (Fig. 1d).

All of the endangered neuronal types share a fifth feature, namely, their long and thin-caliber axons do not undergo myelination or have only a thin myelin sheath (Fig. 1d). Projection neurons attain functional maturity only after their axons have undergone myelination. Neocortical projection neurons in prefrontal or sensory high-order association areas that commence myelination late in life are especially predisposed to the inclusion body pathologies associated with sPD and Alzheimer's disease. Highly susceptible subcortical nuclei include components of the magnocellular nuclei of the basal forebrain, the hypothalamic tuberomamillary nucleus, pedunculopontine tegmental nucleus, substantia nigra, the nuclei of the adjoining ventral tegmental area, coeruleus–subcoeruleus complex, and the nuclei of the raphe system—all of which generate long unmyelinated or sparsely myelinated and slender axons. The diffusely projecting neurons of thalamic intralaminar nuclei also have thinly myelinated axons that extend in a nonspecific manner throughout more than one cortical area. Neurons of the thalamic midline nuclei generate slender axons that furnish thalamo-allocortical circuits and projections to the ventral striatum. All of these thin-axoned thalamic projection cells are prone to develop Lewy pathology. The reverse also applies: Cortical or subcortical projection neurons with a long, sturdy, and heavily myelinated axon are not susceptible (Fig. 1c). Pertinent examples are the Betz cells in the primary motor area and Meynert's pyramidal cells in the striate area (Fallon and Loughlin 1987; Saper 1987; van der Knaap and Valk 1995; Brady et al. 1999; Rüb et al. 2002; Braak and Del Tredici 2004, 2005, 2008).

The entire cerebellum and the precerebellar nuclei are, to a very great extent, also resistant. Subcortical nuclei that do not become involved include the vestibular nuclei, the premotor and motor neurons of the spinal cord and medulla oblongata, the red nucleus, subthalamic nucleus, the dorsal column nuclei, the principal sensory nucleus of the trigeminal nerve, the superior olivary nucleus, and all of the relay nuclei of the thalamus. In contrast to the intralaminar and midline nuclei, the specific thalamic relay nuclei send heavily myelinated projections that form columnar arborizations in layers II–V of defined neocortical areas. Whereas the intralaminar and midline nuclei become severely affected in the course of sPD (Fig. 19), the relay nuclei remain intact. The vagal nerve also serves to illustrate this phenomenon: Its long axons that leave the dorsal motor nucleus and innervate the enteric nervous system do not undergo myelination. By contrast, all of the special motor fibers of the vagal nerve that originate from the ambiguus nucleus and directly innervate striated musculature develop a thick myelin sheath. In sPD, the Lewy pathology appears only in the dorsal motor nucleus and not in the ambiguus nucleus (Fig. 2).

At least four assets or potentially neuroprotective properties characterize a well-developed myelin sheath. (1) Whereas the velocity of axonal conduction increases with growing thickness of the myelin sheath, (2) the metabolic demands placed on the parent nerve cell for the transmission of impulses probably decrease (Fig. 1c). By contrast, rapid-firing projection neurons with unmyelinated or incompletely myelinated axons are subjected to higher energy turnovers and thereby chronically

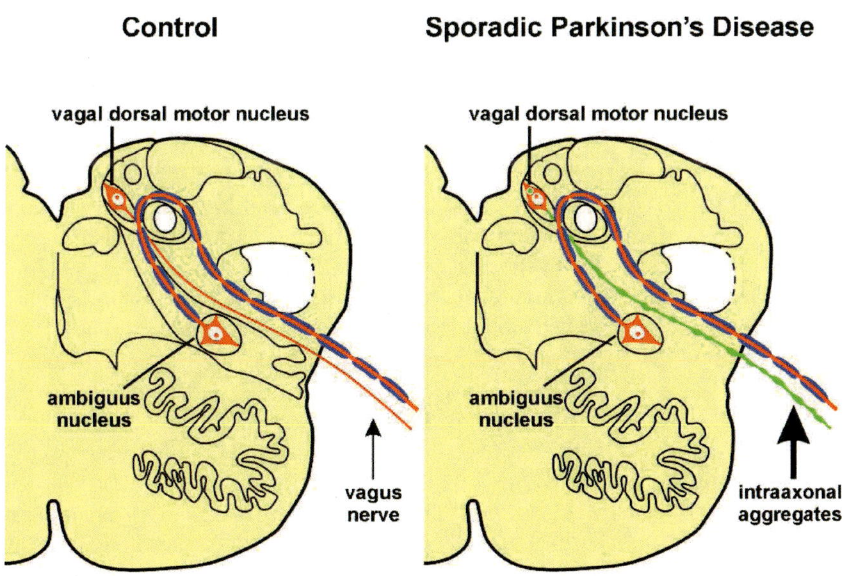

Fig. 2 The vagal nerve serves to illustrate the selective vulnerability shown in Fig. 1b–d. The long axons originating from the preganglionic nerve cells of the dorsal motor nucleus that innervate the enteric nervous system remain unmyelinated for life (*smaller arrow at left*). In contrast, all of the special motor fibers of the vagus nerve that directly innervate skeletal muscle originate from the ambiguus nucleus and develop a thick myelin sheath. In sporadic Parkinson's disease, Lewy pathology (*larger arrow at right*) occurs only in the nerve cells and axons of the vagal dorsal motor nucleus and not in the ambiguus nucleus

exposed to the influences of oxidative stressors that may contribute to the pathogenesis of neurodegenerative brain lesions (Fig. 1d; Hill 1987; Olanow 1992; Beal 1995; Jenner 2003; Jenner and Olanow 1996). (3) A heavily myelinated projection neuron probably possesses a greater degree of stability because the oligodendrocytes that produce and sustain the myelin sheath may also be capable of reducing the axon's propensity for pathological sprouting. (4) Finally, the myelin sheath functions as a protective barrier against viruses and other pathogens. These properties are all the more pronounced the earlier axonal myelination begins and the thicker the myelin sheath becomes throughout the differentiation and maturation process. Seen against this background, projection neurons with a long, late-myelinating (or unmyelinated), small-diameter axon constitute a *locus minoris resistentiae* in the human nervous system. The development of new therapeutic strategies for sPD should be based on the most recent findings regarding the neuronal types that are, or may be, innately capable of withstanding the pathological process and the reasons why such nerve cells are resistant (Kapfhammer and Schwab 1994; Braak and Del Tredici 2004; Del Tredici and Braak 2004).

2 Morphology of Lewy Pathology

2.1 The Protein α-Synuclein

Lewy pathology chiefly consists of misfolded and aggregated forms of the presynaptic neuronal protein α-synuclein that belongs to a larger and highly conserved family of proteins, including β-synuclein and γ-synuclein or breast cancer-specific gene 1 (BCSG1 or persyn). In its natively unfolded state, α-synuclein ranges from 127 to 140 amino acids in length and is soluble in cytosol. The protein contains a lipid binding repeat region followed by a hydrophobic area—the non-Aβ component (NAC) region between residues 61 and 95—and a negatively charged carboxy terminal (Fig. 1e). α-Synuclein can be phosphorylated, nitrated, and is transported by axoplasmic flow (Jensen and Gai 2001; Anderson et al. 2006). It may be involved in the transport of synaptic vesicles and it probably also is capable of microtubule-polymerizing activity. Additional functions are currently unknown (Wakabayashi et al. 1992, 1998; Spillantini et al. 1997; Baba et al. 1998; Clayton and George 1998, 1999; Lavedan 1998; Trojanowski and Lee 1998, 2000, 2003; Golbe 1999; Golbe and Mouradian 2004; Duda et al. 2000; Goldberg and Lansbury 2000; Giasson et al. 2000a, 2001; Galvin et al. 2001; Goedert 2001; Walker and LeVine 2001; Fujiwara et al. 2002; Alim et al. 2004; Norris et al. 2004; von Bohlen und Halbach 2004; Chua and Tang 2006; Gitler and Shorter 2007; Yu et al. 2007; Klegeris et al. 2008; McGeer and McGeer 2008).

Since α-synuclein exists in many, but not all, nerve cell types in the human nervous system, it can be inferred that all of the vulnerable neurons in sPD require sufficient amounts of the natively unfolded protein to become involved (Braak et al. 2001).

On the other hand, not all projection cells with large supplies of α-synuclein are vulnerable, and some of them (e.g., nerve cells in the dorsal tegmental nucleus of Gudden and principal neurons of the inferior olive) withstand the disease process. In mature nerve cells, α-synuclein is primarily located in the axon and its presynaptic terminals, where it mostly binds to the amorphous matrix beneath the regions of abundant synaptic vesicles. Antibodies directed against α-synuclein label both the native (Fig. 6b) and the aggregated protein when the natively soluble protein is precipitated by the fixation procedure. In adequately fixed autopsy material, intact α-synuclein-containing synaptic terminals appear as tightly packed and moderately immunoreactive punctae in the nervous systems of sPD cases and controls alike (Spillantini et al. 1997; Dickson 1999, 2001; Perrin et al. 2000; Goedert 2001; Jensen and Gai 2001; Shults 2006).

2.2 Protein Misfolding and Abnormal Aggregation of α-Synuclein

Under conditions that are still the object of ongoing research, monomeric α-synuclein molecules forfeit part of their binding capacities and undergo a change of conformation by assuming a β-sheet structure (Giasson et al. 2001; Cole and Murphy 2002; Gregersen 2006). Structural motifs, including the hydrophobic stretch following the lipid binding domain of the repeats and encompassing the NAC sequence, probably are critical to this misfolding process (Biere et al. 2000). Misfolded α-synuclein displays a strong affinity to aggregate and, together with other similarly misfolded α-synuclein molecules, undergoes transformation into nearly insoluble inclusions. The intraaxonal and intraneuronal Lewy pathology contains, in addition to α-synuclein, varying combinations of additional proteins, some of which may be induced as a cellular response to inclusion body formation, whereas others may become "trapped" within the inclusions while they are forming (Trojanowski and Lee 1998, 2000, 2002, 2003; Dickson 1999, 2001; Duda et al. 2000; Kopito 2000; Münch et al. 2000; Wakabayashi et al. 2000a; Chung et al. 2001; Goedert 2001; Goedert et al. 2001; Uversky et al. 2001; Walker and LeVine 2001; McNaught and Jenner 2001; McNaught et al. 2002; Ciechanover and Brundin 2003; Eriksen et al. 2003, 2005; Olanow et al. 2004; Olanow and McNaught 2006; Tofaris and Spillantini 2005; Alegre-Abarrategui et al. 2007; Leverenz et al. 2007; Moran et al. 2007; Wakabayashi et al. 2007).

Most of the intrinsic and extrinsic factors that contribute to the induction and maintenance of protein misfolding and aggregation are currently unknown. Similarly, it is not at all clear why these events occur in the predisposed neuronal types described above (Fig. 1a,d) and why such nerve cells do not eliminate the misfolded material by means of rapid ubiquitination and subsequent proteasomal recycling, thus preventing the aggregation process altogether. Apparently, after aggregation, the debris cannot be disposed of by physiological clearance mechanisms in a timely enough manner and the proteinaceous collectivities persist in the form of light microscopically visible inclusion bodies (Ding and Keller 2001; Chung et al. 2001; McNaught and Jenner 2001; Stolzing and Grune 2001).

There is evidence that neither the physiological α-synuclein monomeric molecules nor the filamentous aggregates exert a directly pathogenic influence on the parent nerve cells but, instead, that noxious effects originate chiefly from intermediate oligomeric products known as protofibrils (Chung et al. 2001; Volles and Lansbury 2003; Glabe and Kayed 2006). Nerve cells and axons that contain aggregated α-synuclein or Lewy pathology can survive for years. Neuronal survival, however, is not equivalent to functional integrity. The question has been raised whether Lewy neurites (LNs) and Lewy bodies (LBs) cause premature neuronal death (Kremer and Bots 1993) and, more recently, whether the inclusions contribute to nerve cell loss or dysfunction at all. On the contrary, it has been argued that they are possibly neuroprotective or nonpathogenic (Parkkinen et al. 2005, 2008; Lee et al. 2006; Halliday and McCann 2008; Meredith et al. 2008). Although marked differences may exist among the various types of vulnerable nerve cells, it has been shown that inclusion-bearing nerve cells forfeit some of their functional capacities long before cell death occurs (Beach et al. 2008). It is also logical to surmise that beyond a certain threshold the intraneuronal inclusions ultimately become detrimental to the organism's health (see also Sect. 2.4; Braak et al. 2006c).

2.3
Light Microscopic Visualization of Lewy Pathology

Lewy pathology appears as spindle-shaped or thread-like LNs within neuronal processes and as granular aggregations (Fig. 3h), faintly contoured pale bodies (Fig. 3i), and/or globular LBs in the cell bodies of vulnerable nerve cells (Fig. 3a–c,i,k). Voluminous LNs or LBs are recognizable in standard paraffin sections stained for general overview with hematoxylin-eosin (H&E), and also can be stained with great clarity by using an advanced silver technique (Fig. 3b; Campbell et al. 1987; Sandmann-Keil et al. 1999). Prior to the discovery of α-synuclein, filiform, thread-like LNs and, by contrast, granular aggregations or small LBs could be visualized most effectively with immunoreactions against ubiquitin, a heat shock protein

Fig. 3 a–n (continued) of the Ammon's horn, 6 μm paraffin section. **d–g** Typical intraaxonal Lewy neurites in α-synuclein immunoreactions. **d, e** Club-shaped and filiform Lewy neurites (100 μm polyethylene glycol-embedded sections). **f, g** Varicose and voluminous, flame-shaped Lewy neurites (6 μm paraffin sections). **h–k** Emergence and formation of Lewy bodies in the substantia nigra, pars compacta, α-synuclein immunoreactions, 6 μm paraffin sections. **h** Particulate (i.e., aggregated) immunoreactive particles within a dopaminergic melanoneuron probably precede Lewy body formation. **i** Pale body (*in background*) and Lewy body (*in foreground*). **k** A healthy dopaminergic melanoneuron (*right*) is flanked (*above left*) by a melanized neuron containing seven Lewy bodies. **l–n** Lewy plaques consist of a β-amyloid "core" and a perimeter of α-synuclein-immunoreactive dystrophic neurites. All three stem from the first sector of the Ammon's horn (hippocampal formation; 6 μm paraffin sections). The α-synuclein immunoreaction shown in **l** is supplemented in **m** and **n** by silver staining of β-amyloid (Campbell Switzer pyridine). *Scale bars*: **a–n** 20 μm

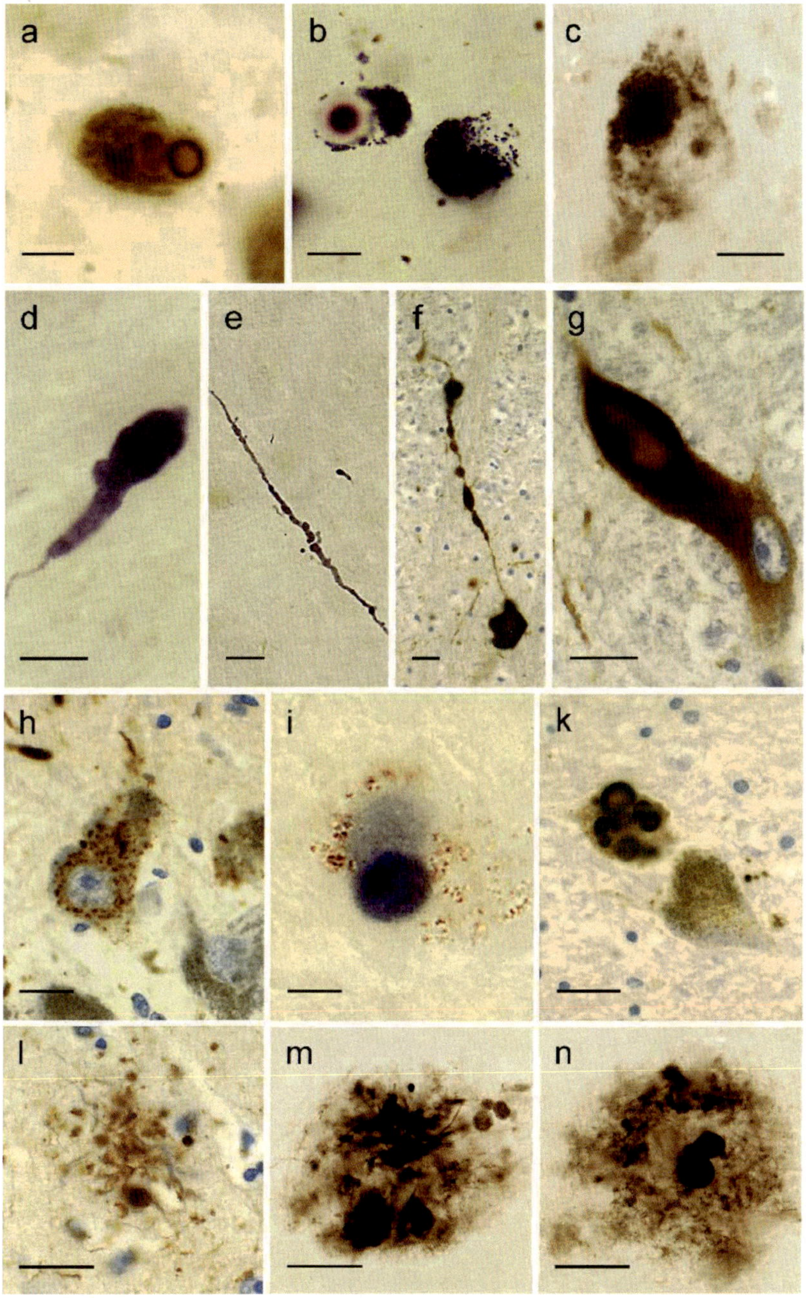

Fig. 3 **a–n** Lewy pathology in sporadic Parkinson's disease as visualized using various methods for demonstrating α-synuclein aggregations. **a–c** Intraneuronal Lewy bodies in immunoreactions against ubiquitin (**a**) and α-synuclein (**c**), as well as in an advanced silver staining technique (Campbell-Switzer pyridine, **b**). **a, b** Projection neurons in the locus coeruleus, 100 μm polyethylene glycol-embedded sections. **c** Mossy cell in the fourth sector

required for the nonlysosomal ATP-dependent breakdown of abnormal proteins (Fig. 3a; Leigh et al. 1989; Lennox et al. 1989). One drawback of ubiquitin immunocytochemistry, however, is costaining of other structures that resemble LBs, such as *corpora amylacea*. Currently, antibodies against α-synuclein (Fig. 3c–l) are the gold standards for recognition of all forms of Lewy inclusions (Cissé et al. 1993; Pollanen et al. 1993; Lowe 1994; Bergeron and Pollanen 1996; Forno 1996; Iwatsubo et al. 1996; Irizarry et al. 1998; Giasson et al. 2000b; Goedert et al. 2001; Takahashi and Wakabayashi 2001, 2005; Jellinger and Mizuno 2003; Braak et al. 2006b; Wakabayashi et al. 2007).

2.3.1 Lewy Neurites

The inclusions in neuronal processes, LNs, have received little attention until comparatively recently (Wakabayashi et al. 1992; Braak et al. 1994, 1999; Saha et al. 2004; Orimo et al. 2005; Mori et al. 2007). Most are located intraaxonally and range from club-shaped, flame-shaped, or branching forms (Fig. 3d–g) to thread-like structures that do not exceed the diameter of a thin axon (Fig. 3e). It is unclear why the formation of LNs precedes, as a general rule, the development of LBs (Braak et al. 2003a, 2006c). Since, in its native state, α-synuclein localizes predominantly to presynaptic terminals, the earliest aggregations could be anticipated to appear there (Kramer and Schulz-Schaeffer 2007). Nevertheless, the first light microscopically detectable accumulations begin proximal to the terminals—in the axon—at least at most sites. The aggregations can fill extensive portions of the axon, leaving free only the presynaptic terminals and the initial axon segment. Terminal axons of corticostriatal and corticothalamic projections appearing as small α-synuclein-immunoreactive dots (see Sect. 8.3) may be exceptions. It remains to be seen whether the abnormal material accumulating within the axon is transported retrogradely to the cell body and contributes there to the formation of LBs (Lewy 1912; Dickson et al. 1991, 1994; Gai et al. 1995; E. Braak et al. 2001a; Duda 2004; Duda et al. 2002; Braak et al. 2007b).

Dense networks of LNs develop mainly in the anterior olfactory nucleus (Fig. 5e,f; Daniel and Hawkes 1992), select subnuclei of the amygdala (Braak et al. 1994; Iseki et al. 1995), and the anterior temporal, insular, subgenual, and cingulate periallo- and proneocortical areas (Kosaka and Iseki 1996; Del Tredici and Braak 2004; Braak and Del Tredici 2008), as well as in the second sector of the Ammon's horn (Fig. 20c–e; Dickson et al. 1991, 1994; de Vos et al. 1995, 1996). LNs that extensively fill out the axon also appear in the white substance, such as the external capsule and medial forebrain bundle. Brainstem thread-like LNs can be seen along the intramedullary and extracerebral route of preganglionic fibers within the vagal nerve (Fig. 7) or in axons of the catecholaminergic fiber tract in the medulla oblongata. Still unresolved is the question why, in sPD, axons of involved dopaminergic projection neurons in the substantia nigra and axons of the pyramidal cells in layers V–VI in the neocortex that furnish corticostriatal and corticothalamic pathways remain nearly free of α-synuclein aggregations, i.e., they display no immunolabeling.

Under the electron microscope, LNs are located in cellular processes that are filled with mitochondria, synaptic vesicles, and dense lamellar bodies (Dickson et al. 1991; Kosaka and Iseki 1996), thereby reinforcing the viewpoint that most LNs are located intraaxonally. In all probability, LNs eventually disrupt the somatopetal/somatofugal transport of substances and, in so doing, impair the functional capacities of involved nerve cells. Impairment of axonal transport also may induce an abnormally high concentration of α-synuclein within the soma and, then, possibly trigger LB-formation within the cell body. At present there are no reports to the effect that the insoluble α-synuclein "backs up" or causes a "traffic jam" at critical junctures within the axon, e.g., branching points or narrow points (but see Saha et al. 2003). Nor can the direction of the transport (somatopetal/somatofugal) be deduced from the shape of the intraaxonal inclusions.

2.3.2 Granular Aggregations

The genesis of LBs generally coincides with the appearance of fine punctate α-synuclein-immunopositive particles in the neuronal soma (Fig. 3h). Such particles tend to be loosely dispersed within deposits of lipofuscin or neuromelanin granules. The pigment granules possibly function as initiation sites for promoting oxidative crosslinking of the abnormal proteinaceous material. This hypothesis is supported by the observation that neither initial traces of immunopositivity nor mature LBs are found amid the patches of Nissl substance and, furthermore, nerve cell types without lipofuscin or neuromelanin granules appear to be resistant to the Lewy pathology (E. Braak et al. 2001a; Kuusisto et al. 2003).

2.3.3 Pale Bodies and Lewy Bodies

Smoothly contoured, pale-staining somatic structures termed "pale bodies" or "glassy degeneration," occur in the somata chiefly of neuromelanin-containing neurons in the substantia nigra (Fig. 3i) and coeruleus-subcoeruleus complex. Pale bodies probably are the precursors of LBs (Dale et al. 1992; Hayashida et al. 1993; Wakabayashi et al. 2007). LBs also are located in the somata of nerve cells and have a rounded or reniform aspect. They are weakly acidophilic structures with smoothly contoured surfaces and are capable of displacing lipofuscin or neuromelanin granules (Fig. 3b). Individual nerve cells can develop multiple LBs (Fig. 3k). Currently there is no evidence that LBs evolve in macroglial cells or non-neuroectodermal cells. Two types of LBs have been described: the classical brainstem type, with an acidophilic hyaline core and a narrow pale-staining halo (Fig. 3b), and a less clearly defined cortical type without a halo (Fig. 3c). Whereas the core of both types is rich in ubiquitin, the halo contains the bulk of α-synuclein (Lewy 1912; Forno and Norville 1976; Gibb and Lees 1988, 1989; Leigh et al. 1989; Irizarry et al. 1998; Saito et al. 2003; Shults 2006).

2.3.4 Lewy Plaques

Lewy plaques (LPs) are spherical structures that consist of core-like extracellular β-amyloid deposits, such as those found in Alzheimer's disease (Fig. 3m,n), and a perimeter of dystrophic α-synuclein immunoreactive LNs (Fig. 3l). They occur in the cerebral cortex and only when cortical LNs and LBs are present. Extracellular β-amyloid precipitations are necessary for the development of LPs, which can attain large diameters, and it is likely that the β-amyloid cores induce the formation of the dystrophic LNs. Clusters of LPs meld together. Other LPs remain isolated within the neuropil in the proximity of blood vessels (Del Tredici and Braak 2004).

2.4 Degradation of Lewy Pathology

The locus coeruleus and substantia nigra undergo neuronal loss soon after the pathological process has reached the pontine and midbrain tegmentum (Figs. 14d and 16f,g). Mature LBs can fill a large proportion of the soma (Fig. 3b,k) and, following cell death, lie free in the neuropil with lipofuscin and neuromelanin granules decorating their surfaces. Under the light microscope, such LBs are no longer surrounded by a neuronal cell body that has a nucleus. Such "extraneuronal" LBs remain visible in the tissue, similar to tombstone tangles in Alzheimer's disease, but they are more rapidly degraded by macrophages than extraneuronal neurofibrillary tangles (Lowe 1994; Kuusisto et al. 2003; Saito et al. 2003; Mikolaenko et al. 2005). In sPD, the sites of the dead nerve cells are marked by the former neuromelanin or lipofuscin contents and remnants of α-synuclein-immunoreactive material.

Deposits of lipofuscin pigment granules are found in most nerve cell types in the human adult (Braak 1980, 1984) and are more difficult for macrophages or astrocytes to degrade than the Lewy inclusions. Lipofuscin and neuromelanin granules are structurally so stable that they remain virtually unaltered by delayed or suboptimal fixation. These assets can be exploited to assess neuronal loss (Braak et al. 2003a, c). Special techniques are available that stain deposits of neuronal lipofuscin granules more prominently than granules of the same type in glial or non-neuroectodermal cells (Braak 1980). Such staining enables the viewer to visualize the breakdown of nerve cells in the form of extraneuronal lipofuscin pigment remains, a phenomenon sometimes described as "pigment incontinence" (Braak et al. 2003a). In the brain of the human adult, neuromelanin is an oxidative byproduct of the biosynthesis of catecholamines and, thus, its presence can be used as a marker for catecholamine-synthesizing nerve cells (Saper and Petito 1982). Since neuromelanin can be labeled specifically (Sandmann-Keil et al. 1999) and its natural color is recognizable even in unstained sections, it is easy to diagnose the degradation of all neuromelanin-laden nerve cell types. Phagocytosing cells take up the neuromelanin and remain in place for relatively long periods of time, thus marking the sites of the lost neurons.

2.5 Lewy Pathology Can Co-occur with Pathology Related to Other Disorders

Whereas Lewy pathology seldom accompanies multiple system atrophy, it co-occurs so frequently with Alzheimer's disease that it has been designated as the "Lewy body variant of Alzheimer's disease" (LBVAD; Hansen 1997). The presence of lesions associated with other proteinopathies (Table 1) does not appear to diminish the pathological influence of LNs and LBs in sPD (Lippa 2003). In fact, there is some evidence for a facilitatory or synergistic effect between the Alzheimer-related lesions and abnormal α-synuclein fibrillization (Lee et al. 2004; Geddes 2005; Pletnikova et al. 2005; Uchikado et al. 2006; Peuralinna et al. 2008).

3 The Evolving Distribution Pattern of Lewy Pathology Associated with sPD Renders Neuropathological Staging Possible

3.1 Incidental Lewy Pathology

Lewy pathology occurs as an incidental finding in nonsymptomatic individuals (Forno 1969). Such LNs and LBs are considered by some to be either natural products of neuronal aging or epiphenomena of other neurobiological processes (Tompkins and Hill 1997; Porta 2002; Jellinger 2004; Saito et al. 2003, 2004; Parkkinen et al. 2005). However, because most individuals examined in nonselected autopsy-based studies typically do not exhibit LNs and LBs, not even at an advanced age (Gibb and Lees 1989; Hughes et al. 1992; Forno 1996; Saito et al. 2004; Chu and Kordower 2007), and since sPD is a dynamic process, incidental lesions can also be viewed as nonbenign entities in a very early phase of the disorder—comparable to the first malignant cells in a carcinoma—that fail to produce clinically detectable symptoms but mark the beginning of a pathological process (Del Tredici et al. 2002; Orimo et al. 2008; Dickson et al. 2008). In two recent studies, incidental cases were shown to have reduced striatal tyrosine hydroxylase (Beach et al. 2008; Dickson et al. 2008), which indicates that even mild pathology may be a precursor of sPD. We regard incidental LNs and LBs as clinically mute but disease-related inclusions and have argued that their existence poses a potential threat to the nervous system because they can result in neuronal dysfunction and nerve cell loss in the course of sPD (Braak and Del Tredici 2008).

3.2 Presymptomatic and Symptomatic Phases

As in other illnesses, some individuals cross the threshold from a presymptomatic or nonsymptomatic phase (Fig. 4e) to the symptomatic manifestation of sPD (Thal et al. 2004). Nonetheless, by the time clinicians make the diagnosis based on initial

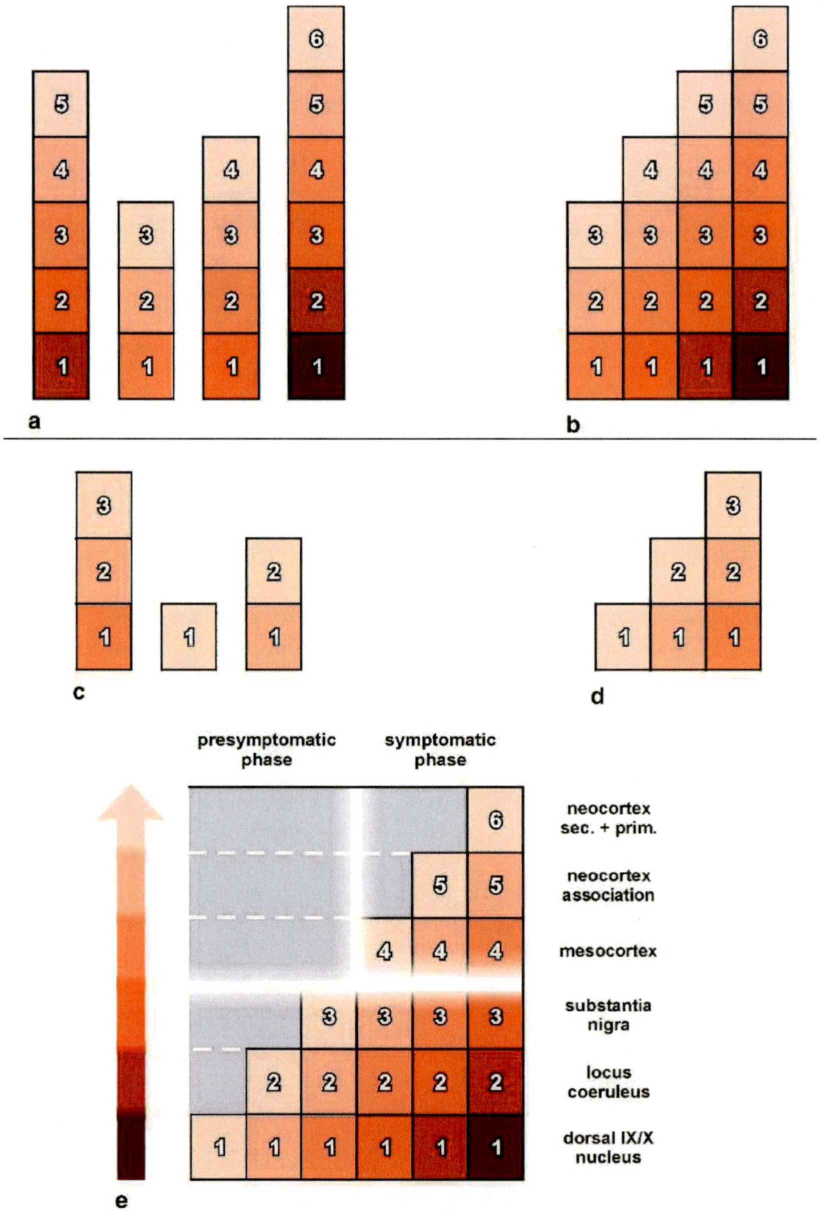
a
b
c
d
presymptomatic phase
symptomatic phase
neocortex sec. + prim.
neocortex association
mesocortex
substantia nigra
locus coeruleus
dorsal IX/X nucleus
e

motor signs, patients are, relatively speaking, in the later phase of a much larger pathological process. The disease "smolders," as it were, in the nervous system for a long time until it has attained such dimensions that dysfunctions become evident. Symptomatic sPD cases can be assigned to one of four neuropathological subgroups that differ from each other with respect to predictable changes in the topographic distribution of Lewy pathology in the brain; each subgroup displays newly affected regions in addition to the lesions at previously involved sites (Fig. 4a). Inasmuch as most autopsy-verified cases of clinically diagnosed sPD can be classified (staged) based on the topographic extent and severity of the Lewy pathology, a sequence of four clinical subgroups has been postulated (Fig. 4b; Braak et al. 2006b).

Relatively little is known about the presymptomatic phase, although it continues to occupy a pivotal position with respect to the pathogenesis of sPD (Litvan et al. 2007). Positron emission tomography (PET) studies have revealed the existence of such a phase in sPD (Sawle 1993; Morrish et al. 1996, 1998; Snow 1996; Brooks 1998, 2000; Brooks et al. 2003). Approximately 5%–20% of nonsymptomatic individuals above the age of 60 display at autopsy mild lesions in regions known to become affected in sPD (Gibb and Lees 1989; Hansen and Galasko 1992; Forno and Langston 1993; Bloch et al. 2006). These cases can be subdivided into three additional subgroups using the same criteria applied above, i.e., lesional distribution pattern and severity (Fig. 4c,d; Del Tredici et al. 2002; Braak et al. 2003a). Changes in topographic distribution provide the logic for arranging all six subgroups such that the progressive nature of the disease process is reproduced by a methodical and predictable sequence of six neuropathological stages (Fig. 4e).

According to this view, the designation "presymptomatic phase" implies that α-synuclein-containing inclusions in persons without classical motor symptoms are the neuropathological equivalent to incipient sPD and the harbingers of the symptomatic phase (Fig. 4d). The pathological process underlying both the presymptomatic and the symptomatic disease phases is marked by the presence of the same types of inclusion bodies in the same susceptible neuronal types in specific regions of the nervous system (Wakabayashi et al. 2007). Insofar as the lesional

Fig. 4 a–e Presymptomatic and symptomatic phases of sporadic Parkinson's disease. **a, b** Nearly all symptomatic cases with Lewy pathology fall into one of four subgroups. Given the consistency of this finding, the four subgroups can be arranged to show disease progression based on topographical distribution pattern and lesional severity or neuronal loss. **c, d** Similarly, nearly all nonsymptomatic cases fall into one of three subgroups that can again be arranged in an ordered sequence. **e** Because it is improbable that all brain sites are affected with identical severity from the very outset, a more cogent explanation would be that mild to moderate lesions develop over time until the threshold from the presymptomatic to symptomatic phase is crossed. *Arabic numerals* represent the six stages of the neuropathological process. It is the topographical distribution and extent of the Lewy pathology via interconnected anatomical fiber tracts as well as neuronal dysfunction or loss at specific sites that provide the basis for the staging system

pattern of the last presymptomatic subgroup closely resembles that of the first symptomatic subgroup, both sets of subgroups combined can be taken to reflect the entire spectrum of the pathological process associated with sPD (Fig. 4e; Del Tredici et al. 2002; Braak et al. 2003a; Neumann et al. 2004; Thal et al. 2004; Takahashi and Wakabayashi 2005; Dickson et al. 2008).

Support for such a staging hypothesis comes from a constellation of nonmotor symptoms that become manifest early and can predate the appearance of classical motor dysfunctions by years. The length of such a presymptomatic phase has been variously estimated to range from 3 to 40 years (Fearnley and Lees 1991; Jenner 1993; Sawle 1993; Morrish et al. 1998; Koller and Montgomery 1997; Koller et al. 1991; Foley and Riederer 1999; Abbott et al. 2001, 2005; Berendse et al. 2001; Doty 2001; Tissingh et al. 2001; Ponsen et al. 2004; Przuntek et al. 2004; Stiasny-Kolster et al. 2005; Wolters and Braak 2006; Wolters et al. 2000; Ross et al. 2006, 2008; Hawkes 2008; Hawkes et al. 1999).

3.3 Vulnerable Regions are Interconnected Anatomically

A wide variety of sites within the nervous system becomes involved in a nonrandom and consecutive manner. The Lewy pathology probably does not evolve simultaneously at all of the susceptible sites but at predisposed locations and progresses in a predictable manner (Fig. 5a). All of the vulnerable regions are anatomically interconnected (Figs. 22 and 29), which indicates that physical contacts between susceptible regions may play a key role in the pathogenesis of sPD (Saper et al. 1987; Pearson 1996; Pearson et al. 1985). In fact, routes exist that permit the propagation of the disease via axonal transport and transsynaptic (transneuronal) transmission of an as yet undetermined but slowly progressive or prion-like pathogen (see also Sect. 4.5; McBride et al. 2001; Braak et al. 2003b; Miwa et al. 2006; Hawkes et al. 2007; Braak and Del Tredici 2008; Lerner and Bagic 2008; Phillips et al. 2008).

Fig. 5 **a** Diagram showing the proposed essentially caudo-rostral pattern of the pathological process (*solid arrows*) associated with the progression of sporadic Parkinson's disease from the lower brainstem (dorsal motor nucleus of the vagal nerve in the medulla oblongata) through the basal midbrain and forebrain and into the cerebral cortex. The severity of the lesions in vulnerable brain regions is rendered by darker degrees of shading (left) that correspond to shading intensity of the lettering (*right*). The perforated arrow is meant to indicate that the disease may begin in the periphery, e.g., in the gastrointestinal tract, and eventually reach the central nervous system transneuronally and via retrograde axonal transport. **b,c** Detail and overview of the anterior olfactory nucleus from a control case, 100 μm-thick polyethylene glycol-embedded section stained with aldehyde fuchsin for lipofuscin pigment and with Darrow red for basophilic material. **d–f** Lewy pathology in the anterior olfactory nucleus. Incidental lesions at neuropathological stages 1 (**d**) and 3 (**e**), and severe lesions seen in a case with stage 6 (**f**) brain pathology. *Scale bar* in **d** is valid for **e** and **f**

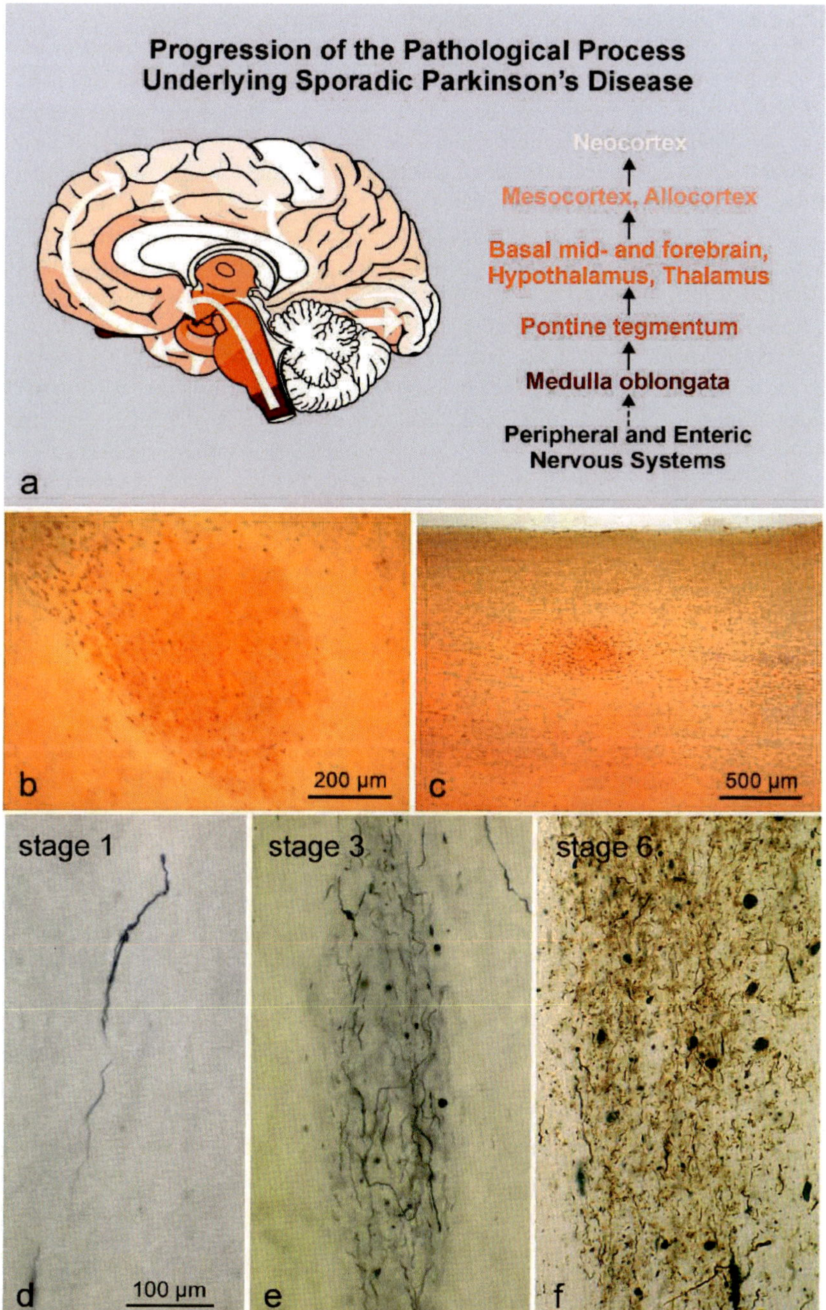
Progression of the Pathological Process
Underlying Sporadic Parkinson's Disease
Neocortex
Mesocortex, Allocortex
Basal mid- and forebrain,
Hypothalamus, Thalamus
Pontine tegmentum
Medulla oblongata
Peripheral and Enteric
Nervous Systems
a
b
200 µm
c
500 µm
stage 1
stage 3
stage 6
d
100 µm
e
f

Although the staging system is not without controversy (Parkkinen et al. 2005; Kalaitzakis et al. 2007; Jellinger 2008), preliminary findings, including those from large autopsy-controlled prospective longitudinal studies, tend to support the hypothesis that sPD begins in the dorsal motor nucleus of the vagal nerve and adjoining intermediate reticular zone, anterior olfactory structures, and perhaps also portions of the spinal cord, PNS, and ENS (Mikolaenko et al. 2005; Braak et al. 2006a; Bloch et al. 2006; Ross et al. 2006; Duda et al. 2007; Dickson et al. 2006; Fumimura et al. 2007; Minguez-Castellanos et al. 2007; Halliday and McCann 2008; Halliday et al. 2006, 2008; Parkkinen et al. 2008). From the periphery, the process may gain access to the lower brainstem via the vagal nerve and then take an essentially ascending path through vulnerable regions of the basal mid- and forebrain until it reaches the cerebral cortex (Table 2; Fig. 5a). The caudo-rostral advance of the pathological process within the brain is another major attribute that permits recognition of each neuropathological stage (Müller et al. 2005; Braak and Del Tredici 2008). No remarkable nerve cell loss takes place in the earliest stages, but whether mild neuronal loss is negligible or proves functionally deleterious may well depend on comorbidities, neuronal reserve, and the nerve cell types involved, as well as lesional distribution pattern. A modest, nonrandom loss of nerve cells probably impairs function, whereas comparable but random neuronal loss is unlikely to do so.

4 Stage 1

The first LNs within the brain are seen at two sites more or less simultaneously, namely the dorsal motor nucleus of the vagal nerve and adjoining intermediate reticular zone as well as the olfactory bulb and anterior olfactory nucleus. The human olfactory bulb and related areas (anterior olfactory nucleus, olfactory stalk, olfactory tubercle, and piriform cortex) are poorly developed. Well-differentiated portions of the olfactory system in the human brain include the periamygdalar cortex, the olfactory-related portion of the entorhinal region, and the medial subnucleus of the amygdala. The septal nuclei and also hypothalamus receive information from the olfactory tract (Price 2004).

4.1 Pathology in Olfactory Structures

LNs and LBs develop in the cellular islands of the anterior olfactory nucleus dispersed throughout the olfactory tract (Fig. 5d–f; Daniel and Hawkes 1992; Pearce et al. 1995; Del Tredici et al. 2002). LNs are visible there in stage 1 cases (Fig. 5d) and subsequently form a dense network (Fig. 5e,f). Involved mitral and tufted cells of the olfactory bulb are subject to thinning owing to neuronal loss. The olfactory epithelium remains devoid of LNs/LBs (Duda et al. 1999). For reasons that are still poorly understood, Lewy pathology develops in more remote olfactory sites from stage 3 onwards without encroaching on nonolfactory cortical areas or subcortical nuclei (Braak et al. 2003a, 2004, 2006b; Del Tredici and Braak 2004; Hubbard et al. 2007). Early involvement of the olfactory system in sPD is congruent with clinical reports of olfactory dysfunctions, frequently prior to the manifestation of motor symptoms (Sakuma et al. 1996; Mesholam et al. 1998; Liberini et al. 2000; Doty 2001, 2008; Doty et al. 1992; Berendse et al. 2001; Tissingh et al. 2001; Hawkes 2003; Hawkes et al. 1997, 1999; Ponsen et al. 2004; Sommer et al. 2004; Stiasny-Kolster et al. 2005; Haehner et al. 2007; Kranick and Duda 2008; Ross et al. 2008; Westermann et al. 2008).

4.2 Basic Organization of the Medullary Autonomic Region

Motor and sensory nuclei of the glossopharyngeal and vagal nerves, including parts of the reticular formation, form an extended complex in the lower brainstem referred to as the medullary autonomic region. Frontal sections display the region as a broad and obliquely downwardly oriented stripe extending from the ala cinerea of the fourth ventricle to a shallow sulcus above the inferior olive (Fig. 6a). The region is part of the reticular formation and can be distinguished from its other portions by the presence of melanized neurons (A1, A2 groups). Special components include the motor nuclei of the glossopharyngeal and vagal nerves, the viscerosensory nuclei of the solitary tract, the gelatinosus nucleus (gel; Fig. 6a), and area postrema. The intermediate reticular zone (irz; Fig. 6a) contains premotor autonomic neurons and forms a cardiorespiratory network that controls input to spinal preganglionic motoneurons (Huang and Paxinos 1995). Despite its heterogeneous components, the medullary autonomic region constitutes a functional and architectonic entity. It has pivotal significance for the regulation of cardiovascular and respiratory functions (Blessing 2004). In the course of sPD, marked pathological changes can be observed in this region, whereas immediately adjoining areas remain virtually intact.

4.3 Involvement of Preganglionic Parasympathetic Projection Neurons

In most autopsy cases, the main starting point of the pathological process in the brain is the dorsal motor nucleus of the vagal nerve (dmX; Figs. 6a and 7a,b). LNs are visible in this nucleus accompanied by faintly immunoreactive α-synuclein aggregations in medullary and peripheral portions of the long, thin, and unmyelinated axons (Figs. 2 and 7b; Del Tredici and Braak 2008) generated by the cholinergic vagal preganglionic parasympathetic projection neurons that connect the CNS with postganglionic neurons of the ENS (Fig. 8a; Huang et al. 1993; Hopkins et al. 1996). Shortly thereafter, LBs develop in the somata of the same cholinergic projection neurons (Lewy 1912; Gai et al. 1992; Del Tredici et al. 2002; Benarroch et al. 2006). Other components of the dorsal vagal area, i.e., the gelatinosus nucleus (gel; Fig. 6a,c), area postrema, and most of the small-celled nuclei surrounding the solitary tract (*sol*; Fig. 6a), are minimally affected or unaffected (Figs. 6a and 8a). Similarly, the multipolar motoneurons of the

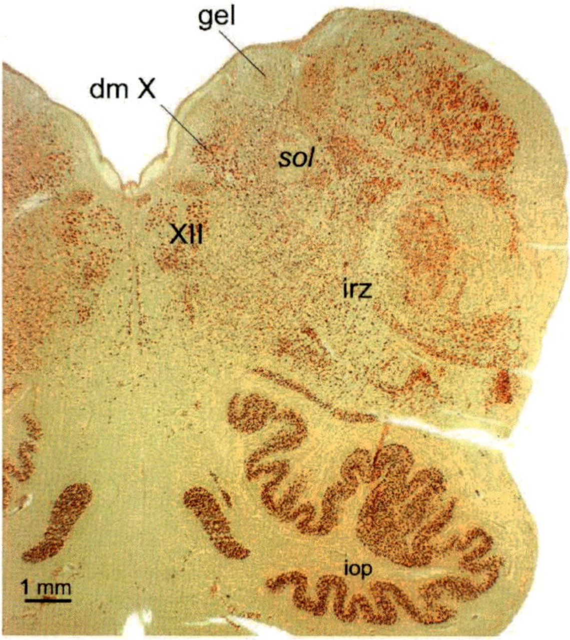

a

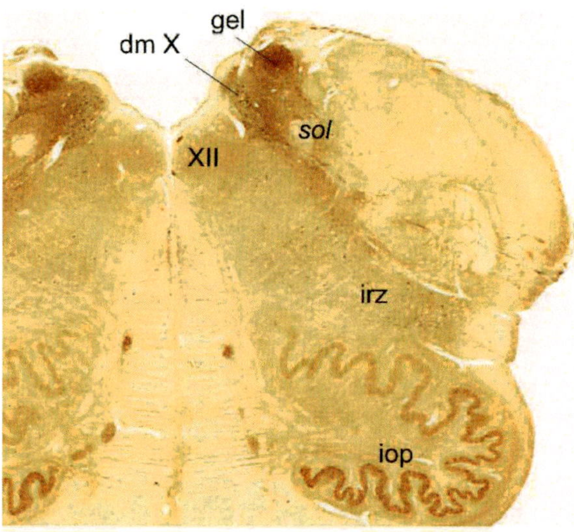

b

Fig. 6 Transverse sections through the lower brainstem (medulla oblongata). **a** Overview of the dorsal motor vagal area (*dm X*) and adjoining intermediate reticular zone (*irz*) from a control case for orientation, 200 µm polyethylene glycol-embedded section, stained with aldehyde fuchsin for lipofuscin pigment and with Darrow red for basophilic material. **Abbreviations:** *gel*, gelatinosus subnucleus; *iop*, principal subnucleus of the inferior olivary nucleus; *sol*, solitary tract; *XII*, motor nucleus of the hypoglossal nerve. **b** Transverse section at about the same latitude as **a** (control case, 100 µm, immunoreaction against the native

ambiguus nucleus together with their thickly myelinated axons that innervate the striated musculature of the larynx and upper esophagus do not become involved at any point in the pathological process (Fig. 8a). The catecholaminergic projection melanized neurons (equivalent to the rodent A2 group) of the motor nucleus of the vagal nerve and those in the intermediate reticular zone (A1 group) do not project to the periphery via the vagus nerve but, instead, generate ascending projections to higher levels of the CNS (Saper et al. 1991). Both groups become involved later, from stage 3 onwards (Braak et al. 2004, 2006b,c).

4.4 Pathology in the Enteric Nervous System

Lewy pathology in the ENS (Costa et al. 2000; Furness 2000; Anlauf et al. 2003; Benarroch 2007; Grundy and Schemann 2007) is known to occur in both nonsymptomatic and symptomatic cases (den Hartog Jager and Bethlem 1960; Forno and Norville 1976; Qualman et al. 1984; Wakabayashi et al. 1988, 1990, 1991, 1993; Wakabayashi and Takahashi 1997b; Takeda et al. 1993). To date, ENS lesions have only been encountered in cases with Lewy pathology beyond the level of the vagal dorsal motor nucleus (i.e., stage 1 and above). It has yet to be shown whether LNs and LBs in the ENS are capable of developing in the absence of the lesions in the CNS (Braak et al. 2006a). Inhibitory VIP motoneurons of the ENS that directly receive vagal preganglionic terminals are prone to develop the lesions (Wakabayashi et al. 1990, 1991; Hornby and Abrahams 2000; Phillips and Powley 2007). Widespread but thinly distributed LNs are found throughout the wall of the gastric cardia, fundus, and pylorus regions, mostly in the fiber strands interconnecting enteric ganglia of the Auerbach plexus (Fig. 9a–d). α-Synuclein-immunoreactive axons from the Meissner plexus are seen reaching into the muscle layer of the mucosa and beyond into the mucosal lamina propria (Fig. 10a) where they take an upward course parallel to the gastric glands (Fig. 10b–d). At this location, the affected axons are only micrometers away from the body's innermost environment (Braak et al. 2006a).

4.5 Is sPD Inducible by a Neurotropic Pathogen Akin to a Slow Virus?

The successive involvement of all of the previously mentioned medullary sites has prompted the question whether a neurotropic pathogen (Braak and Del Tredici 2008; Braak et al. 2003b) induces protein misfolding and α-synuclein aggregation in neurons of the upper gastrointestinal tract, and progresses by way of retrograde

Fig. 6 (continued) α-synuclein). Note that the dorsal motor nucleus of the vagal nerve and, in particular, the gelatinosus subnucleus contain large amounts of the native (soluble) protein. Just as remarkably, nuclei that usually are resistant to Lewy pathology may have rich stores of unfolded α-synuclein, such as the inferior olivary nucleus (note the gradient between the superior and inferior lamellae). *Scale bar* in **a** also applies to **b**

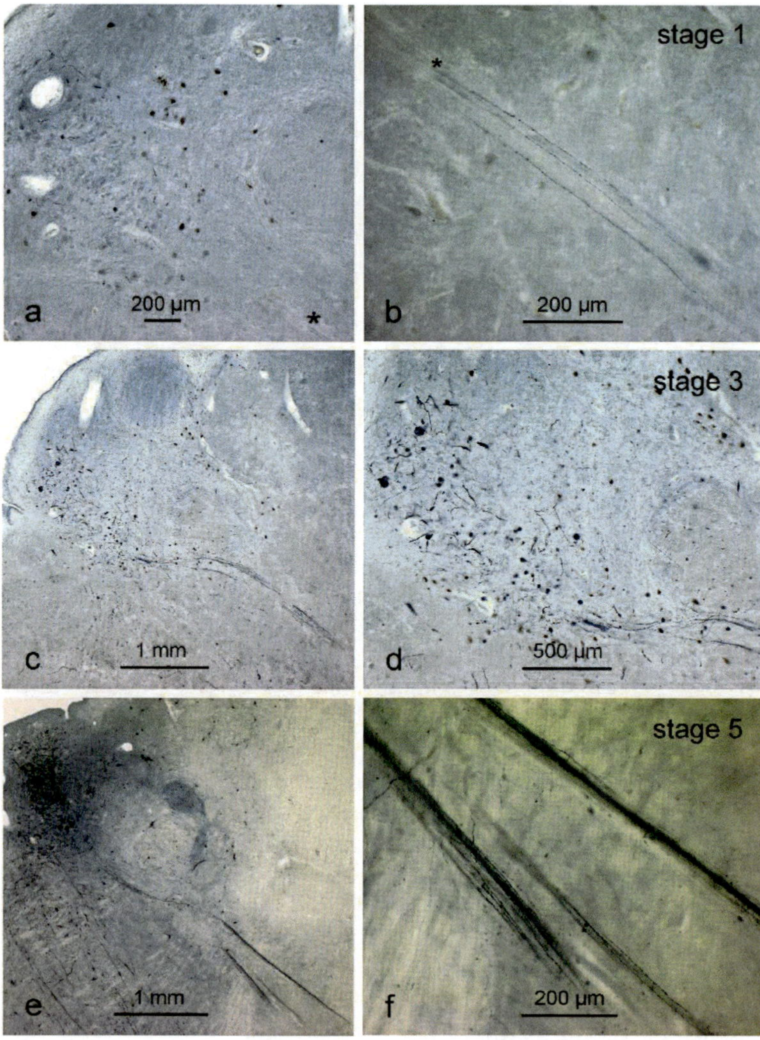

Fig. 7 a–f Increasing severity of the lesions in the dorsal motor nucleus of the vagal nerve in stages 1, 3, and 5. **a, b** The pathology often commences with only a few Lewy neurites, as seen upon closer inspection in **a**. The dark spots correspond to still uninvolved melanized neurons. *Asterisks* in **a** and **b** point to affected axons of the preganglionic vagal nerve cells that contain thread-like α-synuclein aggregations. **c, d** Worsening Lewy pathology in a stage 3 case. Note that the gelatinosus subnucleus (**c**), which consistently contains a large amount of α-synuclein (see Fig. 6b), remains virtually unaffected in sporadic Parkinson's disease. **e, f** Severe involvement of the dorsal motor vagal nucleus at stage 5. Axons of the preganglionic nerve cells are filled with aggregated material that, nevertheless, does not produce conspicuous swellings or torpedo-like inclusions but is evenly dispersed throughout the axon

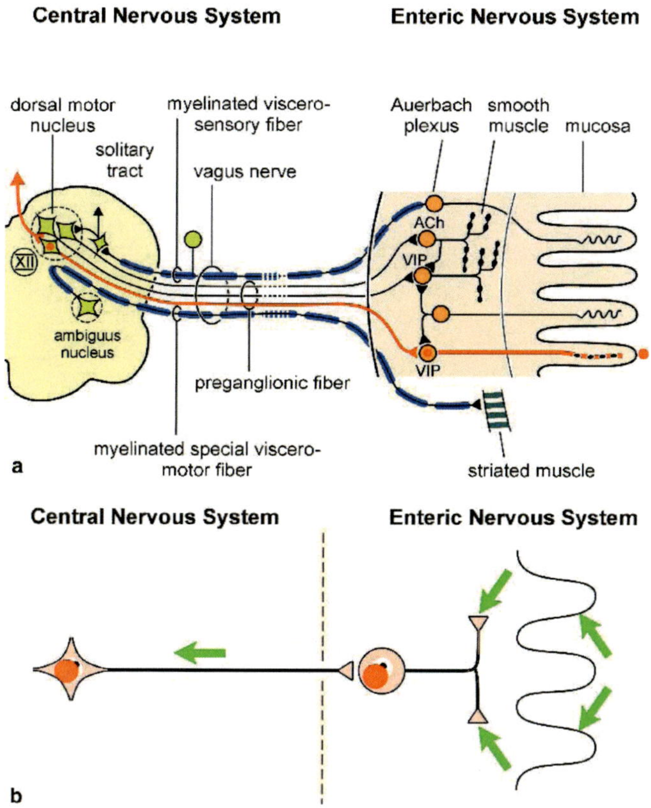

Fig. 8 **a** Diagram of the vagus nerve connecting the enteric nervous system with the lower brainstem (medulla oblongata) of the central nervous system. Myelinated viscerosensory fibers from the gut terminate in the small-celled nuclei that surround the solitary tract. Myelinated visceromotor fibers from the ambiguus nucleus innervate the striated musculature of the upper esophagus. Unmyelinated preganglionic fibers originating from the dorsal motor nucleus of the vagal nerve terminate on ganglion cells of Auerbach's plexus, which innervate the smooth muscle of the gut wall. Lewy pathology develops in the unmyelinated preganglionic axons of the vagal nerve, whereas the axons of the ambiguus nuclei and those surrounding the solitary tract do not become involved in sporadic Parkinson's disease. **Abbreviations:** *ACh*, acetylcholine, cholinergic neurons; *VIP*, vasoactive intestinal peptide, VIPergic neurons. **b** The fact that terminal axons of the Meissner plexus are capable of developing α-synuclein aggregations early in the disease process has given impetus to the search for an environmental neurotropic pathogen that might induce the formation of the inclusion body pathology in sporadic Parkinson's disease. The hypothetical pathogen would transgress the epithelial lining (*arrows at far right*), penetrate into the synapses or axons, and induce the process of protein misfolding and aggregation. The axons are not encased by a protective myelin sheath and are practically defenseless against such an assault. Once the first enteric nerve cells have become involved, the pathogen would be transported transneuronally to the next neuron. The most likely link between the enteric and central nervous system is the preganglionic fibers of the vagus nerve. In this manner, a prion-like pathogen could overcome the considerable distance from the enteric mucosa to the central nervous system via retrograde axonal transport (*arrow at left*). Under physiological conditions, neurotrophic factors are taken up from the axonal environment and transferred retrogradely to the cell body. Viruses, however, also utilize this pathway to gain access to the nervous system

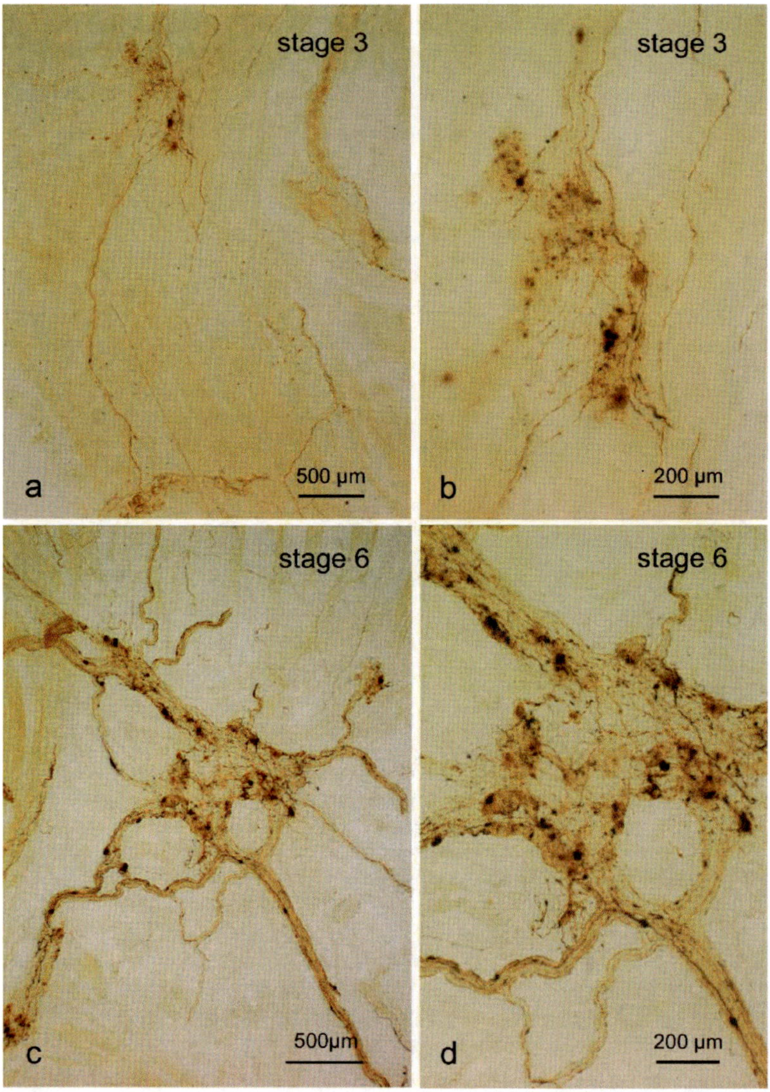

Fig. 9 a–d Lewy pathology in the gastric Auerbach plexus seen in immunoreactions against α-synuclein; 100 µm sections cut tangentially to the tissue surface (syn-1, Transduction and BD Bioscience Laboratories). **a, b** An intramural ganglion in a section cut tangentially to the surface of the gastric cardia from an asymptomatic individual with stage 3 brain pathology. Note that, even at this early stage, α-synuclein aggregations almost completely fill individual axons within the fiber bundles interconnecting the ganglia. To date, screening of a large series of nonselected autopsy cases reveals that this pathology is present in individuals with at least stage 1 brain pathology. Micrograph in **b** shows a detail from **a**. **c, d** In late-stage cases (here stage 6), heavy involvement of the enteric nervous system is one of the reasons why so many patients experience considerable gastrointestinal dysfunction. Micrograph in **d** shows a detail from **c** at higher magnification

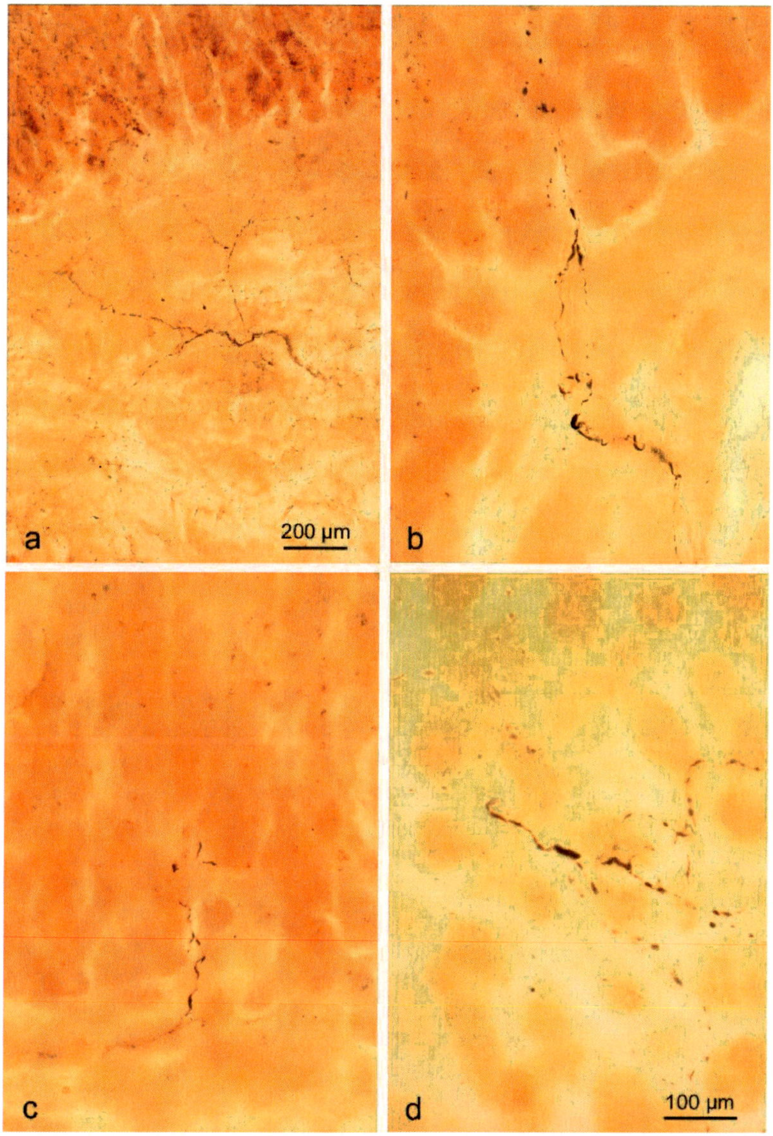

Fig. 10 a–d Lewy pathology in the gastric Meissner plexus seen in immunoreactions against α-synuclein; 100 µm sections cut perpendicularly (**a–c**) or tangentially (**d**) to the tissue surface (syn-1, Transduction and BD Bioscience Laboratories). **a** Ramifications of abnormally altered axons permeate the submucous layer. **b–d** Abnormal axons penetrate the muscularis mucosa to reach the lamina propria of the mucosa where they run parallel to the gastric glands (*above left* in **b**) and ramify there (micrographs from a case with stage 6 brain pathology). Remarkably, the Lewy pathology exists only micrometers away from the epithelial surface of the stomach and esophagus. *Scale bar* in **d** is valid for **b** and **c**

axonal transport and transsynaptic transmission into the CNS (Fig. 8b; Hawkes et al. 2007; Phillips et al. 2008). Epidemiological studies indicate that prolonged exposure to chemicals (e.g., rotenone) is associated with an increased risk of sPD, whereas a reduced risk is mentioned in connection with chronic nicotine abuse (Morens et al. 1996; Hellenbrand et al. 1997; Hernán et al. 2002; Tanner et al. 2002). Both agents act upon the viscous nasal mucous, and swallowed nasal secretions come into direct contact with the epithelia of the upper gastrointestinal tract (Baker and Genter 2003; Hawkes et al. 2007). Probably, exogenous and endogenous factors alike work together to induce and modify the misfolding process; putative causative agents are not limited to slow neurotoxins but include endogenous proteins, such as those contained in nasal mucous (Casado et al. 2005). A truncated protein in β-sheet form might be capable of penetrating the epithelial lining and be taken up by unmyelinated axons. Uptake of exogenous substances is known to occur preferentially at synapses. From there, transfer of the material to the soma takes place via retrograde axonal transport (Sotelo and Triller 1997). Neuroactive substances—including neurotropic viruses, unconventional pathogens with prion-like properties, or slow neurotoxins—are taken up in this manner, e.g., by receptor-mediated endocytosis (Strack et al. 1989; Morrison et al. 1991; Sabin 1996; Card 1998; Helke et al. 1998; Rinaman et al. 2000; McBride et al. 2001; Nicotera 2001; Palka-Santini et al. 2003; Matsuda et al. 2004; Miwa et al. 2006). As pointed out previously (see Sect. 1.2), the myelin sheath serves as a first line of defense against viruses attempting to pass from the surroundings into the axon (Hill 1987). Inasmuch as the vulnerable cells of the ENS and the preganglionic neurons of the dorsal motor vagal nucleus do not possess a protective myelin sheath, viruses or other neuroactive pathogens could easily gain entry. In this context, it is worth reiterating that the vagal lesions only develop in unmyelinated preganglionic motor neurons, whereas the sturdily myelinated neurons that innervate the striated musculature are spared (Figs. 2 and 8a).

Most neuronal types located within the CNS are protected against uptake of substances from the extracellular milieu beyond the CNS by the blood–brain barrier. Only axons of nerve cells that project outside the CNS, such as those given off from preganglionic neurons of the dorsal motor vagal nucleus, lack this protection. An intravenous injection of horseradish peroxidase in the rat, for instance, results in retrograde labeling of the dorsal motor vagal nucleus (Broadwell and Brightman 1976; Sabin 1996). Provided the putative pathogen were to be capable of passing through the gastric mucosal barrier, it could well be taken up by terminal axons of susceptible postganglionic visceromotor neurons and pass into unmyelinated preganglionic fibers of the vagus nerve. In this manner, the postulated pathogen could utilize the vagus nerve as a vehicle to overcome the considerable distance from the gastric mucosa to the CNS by retrograde axonal transport (Fig. 8a,b). Such a hypothesis offers the most economical explanation both for the caudorostral disease propagation that characterizes the sPD neuropathologically and its comparatively long premotor phase (Hawkes et al. 2007). We suggested that the gastric mucosa appears to be among the more likely candidates for such an assault because the bulk of its innervation originates from the vagal dorsal motor nucleus

(Holst et al. 1997). Moreover, ingested foods and the chyme linger in the stomach where the thin mucous membrane is susceptible to microbleeding and chronic infection, e.g., *Helicobacter pylori* (Strang 1965; Beró et al. 1992; Altschuler 1996; Konturek et al. 1998; Dobbs et al. 2000; Palka-Santini et al. 2003).

4.6 Involvement of Preganglionic and Postganglionic Sympathetic Projection Neurons

Functions of the gastrointestinal tract are predominantly controlled by parasympathetic vagal output (Travagli et al. 2006). However, this parasympathetic influence is constantly modified and, as needed, inhibited by the influence of the sympathetic antagonist. Sympathetic relay centers include the intermediomedial and intermediolateral nuclei of the spinal cord and peripheral ganglia, such as the coeliac ganglion (Loewy 1990; Sun 1995; Jänig 1996). Lewy pathology in preganglionic and postganglionic sympathetic neurons has been observed in the presymptomatic phase of the disease (Wakabayashi and Takahashi 1997a; Klos et al. 2006; Bloch et al. 2006; Braak et al. 2007a). Figure 11 shows the advanced pathology observable there in a symptomatic patient. Given the difficulties associated with obtaining and processing the spinal cord at autopsy, representative segments of the spinal cord in sPD have been investigated in only three independent studies to date (Bloch et al. 2006; Klos et al. 2006; Braak et al. 2007a). All of the groups observed Lewy pathology in the spinal cords of individuals in the presymptomatic disease phase with the equivalent of stage 2–3 lesions.

Obstipation is an early symptom that precedes the typical somatomotor symptoms by years (Abbott et al. 2001, 2005, 2007). Modern noninvasive diagnostic instruments, such as stabile isotope breath test, three-dimensional ultrasound, and gastrointestinal functional magnetic resonance imaging for global assessment of human gastrointestinal functions, may result in improved diagnosis of constipation, dysfunctional gastric emptying, and other nondopaminergic aspects associated with early sPD (Hardoff et al. 2001; Müller et al. 2001; Jost and Eckardt 2003; Goetze et al. 2005; Schwizer et al. 2006; Gilja et al. 2007; Verbaan et al. 2007). The cardiac sympathetic denervation that also occurs early in sPD can be assessed with the aid of ^{1}25I-metaiodobenzylguanidin scintigraphy (Goldstein et al. 2000; Takatsu et al. 2000; Li et al. 2002; Sharabi et al. 2003; Taki et al. 2004), and a corresponding neuropathological substrate for the clinical finding has not only been found but has also been verified in a series of incidental cases (Iwanaga et al. 1999; Amino et al. 2005; Orimo et al. 2005, 2007; Fujishiro et al. 2008).

4.7 Pathology in Lamina I of the Spinal Cord

Figure 12a,b depicts LNs and LBs in lamina I (pain system) of the thoracic spinal cord from a sPD patient at stage 3. Some of the neurons there are almost totally filled with aggregated α-synuclein (Fig. 12b). LNs extend from lamina I to the intermediate

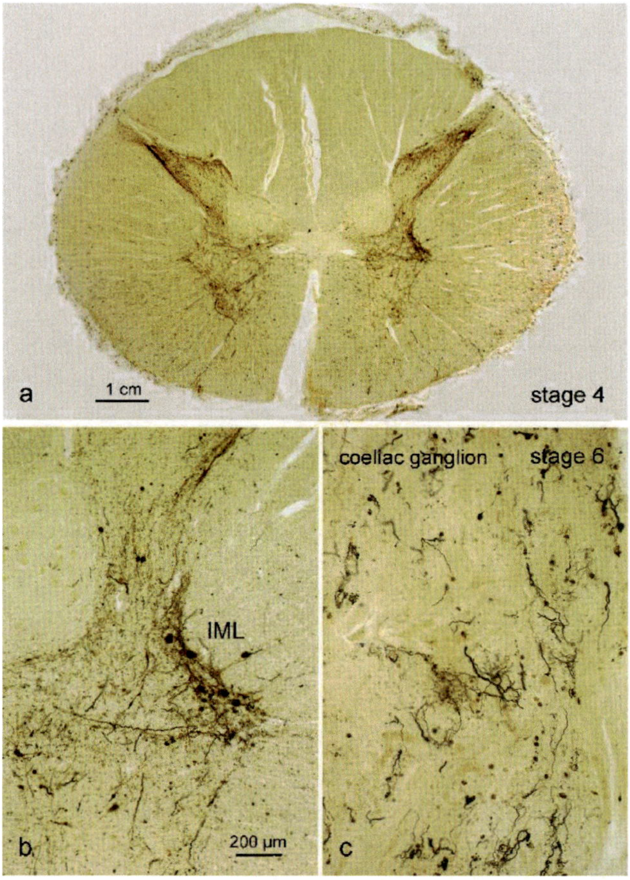

Fig. 11 a, b Lewy pathology in the spinal cord. **a** Overview of the twelfth thoracic segment at stage 4. **b** Detail micrograph of **a** showing involvement of the intermediolateral nucleus (*IML*). Several multipolar sympathetic preganglionic projection neurons are filled with α-synuclein aggregates. By contrast, the dorsal nucleus (pale rounded area at *upper left* representing Clarke's column) is virtually uninvolved. **c** Lewy neurites and Lewy bodies are distributed throughout the entire coeliac ganglion at stage 6. Immunoreactions against α-synuclein, 100 µm. *Scale bar* in **b** is valid for **c**

horn where they come close to the sympathetic relay nuclei. By contrast, lamina I of the spinal trigeminal nucleus, lamina II and the following layers of the spinal cord, and the nerve cells of spinal ganglia that provide input to lamina I remain nearly free of Lewy pathology. From the lowermost cervical segments, the meshwork of LNs in lamina I increases gradually in density further caudally and reaches its culmination in the lower thoracic, lumbar, and sacral segments. LNs also occur in the commissures below the central canal and in the ventrolateral funiculus (Braak et al. 2007a).

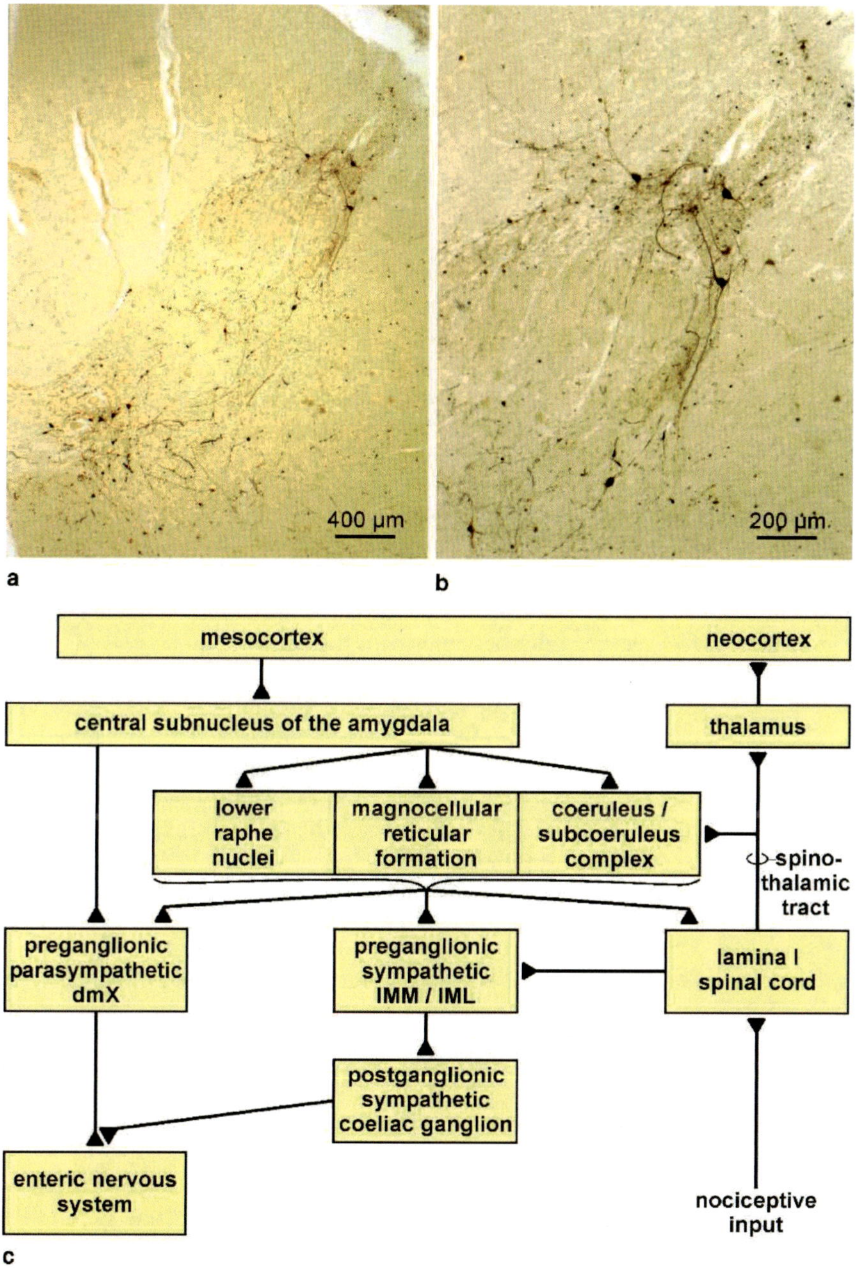

Fig. 12 **a, b** Lewy pathology in the dorsal horn at the level of the tenth thoracic segment at stage 5. The involvement of relatively large lamina I multipolar neurons predominates. The α-synuclein aggregates almost completely fill the somatodendritic domain of these cells. Immunoreactions against α-synuclein, 100 µm. **c** Diagram of the main connections between the enteric nervous system, sympathetic and parasympathetic relay nuclei that influence the ENS, and lamina I of the spinal cord. Note that noxious stimuli have an excitatory influence on sympathetic outflow. **Abbreviations:** *dmX*, dorsal motor nucleus of the vagal nerve, *IMM/IML*, intermediomedial and intermediolateral nuclei of the spinal cord

The diagram in Fig. 12c summarizes the known pathways between lamina I and the autonomic centers of the lower brainstem and spinal cord. Unmyelinated and sparsely myelinated primary afferent Aδ and C fibers in the periphery transfer thermal and nociceptive stimuli from all parts of the body to the CNS. These fibers accumulate at the tip of lamina I and synapse almost exclusively with lamina I projection neurons there. The same nerve cells also receive modulatory supraspinal input from the lower raphe nuclei, reticular formation, coeruleus–subcoeruleus complex, and hypothalamic paraventricular nucleus—sources that generate projections both to the sympathetic preganglionic projection neurons of the spinal cord and to the parasympathetic preganglionic neurons of the vagal dorsal motor nucleus (Fig. 12c). Axons of lamina I neurons partially cross the midline, provide excitatory input to sympathetic preganglionic neurons, and then ascend in the ventrolateral funiculus as components of the spinothalamic tract (Craig 1996, 2003; Willis and Westlund 1997). In addition, they exert an excitatory influence upon the coeruleus–subcoeruleus complex (A6, A7 groups; Fig. 12c). In this context, it should not be forgotten that noxious stimuli have an excitatory effect on sympathetic outflow (Loewy 1991; Tracey 2005; Benarroch 2006).

From stage 4 onwards, Lewy pathology develops in the intralaminar nuclei of the thalamus (Fig. 19a–c), which, together with the periaqueductal gray and anterior cingulate areas, belong to the medial pain system (Bentivoglio et al. 1993; Vogt et al. 1993; Vogt and Sikes 2000). Complaints of painful sensations are not infrequent in early sPD and, combined with other signs including olfactory and autonomic dysfunction, pain can be a precursor of the more typical motor symptoms (Ford 1998; Waseem and Gwinn-Hardy 2001; Buzas and Max 2004; Djaldetti et al. 2004; Sage 2004; Scherder et al. 2005). It is uncertain whether an early reduction of excitatory effects on sympathetic outflow owing to noxious stimuli could be utilized for diagnostic purposes. Such effects, for instance, could be evaluated with the aid of the cold pressor test (Hilz et al. 2002; Wirch et al. 2006).

5 Stage 2

5.1 Involvement of the Medulla Oblongata and Pontine Tegmentum

In stage 2 the inclusion body pathology remains confined to the medulla oblongata and pontine tegmentum. The lesions in the dorsal motor nucleus of the vagal nerve and intermediate reticular zone increase (Fig. 22) and the disease process passes beyond the limits of these nuclei to include: (1) portions of the lower raphe nuclei, particularly the great raphe nucleus (rm; Fig. 13a,b, and c–e), (2) magnocellular portions of the reticular formation (gigantocellular nucleus, gig; Fig. 13a), and (3) catecholaminergic melanized neurons of the coeruleus–subcoeruleus complex (equivalent to the rodent A6 and A7 groups) (Table 2; Fig. 14a and b–d). The appearance of LNs precedes the development of LBs in each of these

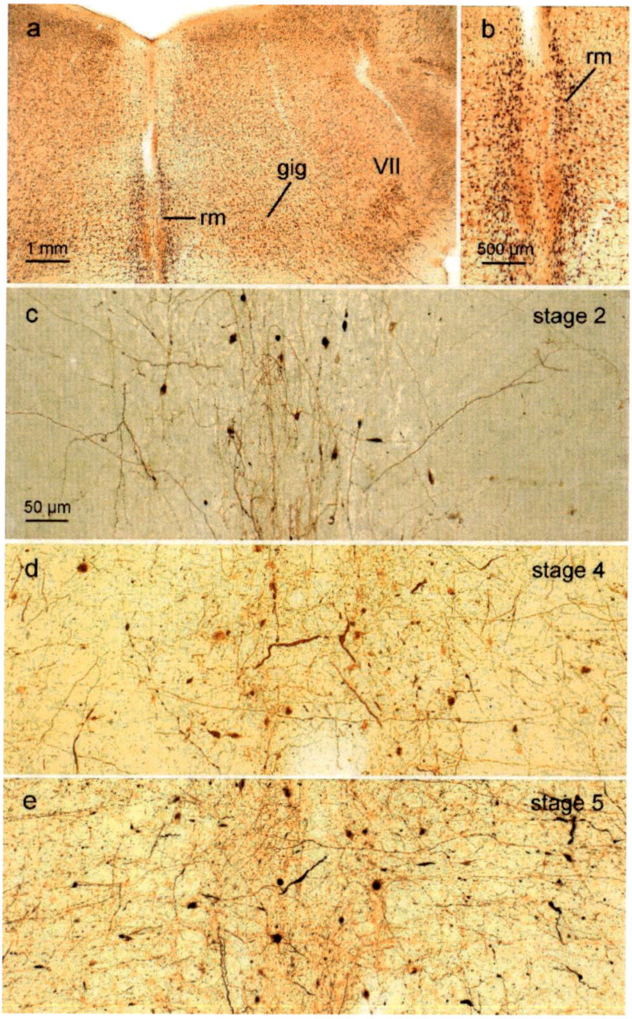

Fig. 13 a–e Lewy pathology in the gain setting lower raphe complex and magnocellular nuclei of the reticular formation. **a, b** Overview and detail micrographs of the great raphe nucleus (*rm*) and gigantocellular reticular nucleus (*gig*) from a control case at the level of the motor nucleus of the facial nerve (*VII*), 100 µm thick polyethylene glycol-embedded section stained with aldehyde fuchsin for lipofuscin pigment and with Darrow red for basophilic material. **c–e** The great raphe nucleus and adjacent portions of the reticular formation in 100 µm thick polyethylene glycol-embedded sections with immunoreactions against α-synuclein at stages 2, 4, and 5, reveal the gradual increase in the severity of the pathology. Scale bar in **c** also applies to **d** and **e**

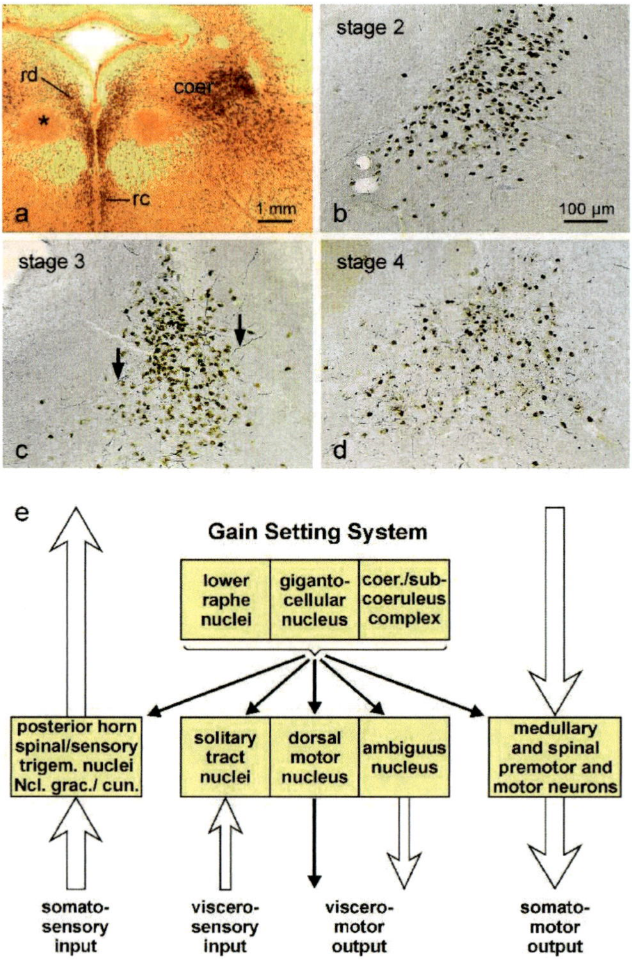

Fig. 14 a–d Lewy pathology in the gain setting coeruleus-subcoeruleus complex (*coer*). **a** Overview of the pontine tegmentum at the level of the dorsal tegmental nucleus of Gudden (*asterisk*) from the same control case as in Fig. 13a (100 µm-thick polyethylene glycol-embedded section stained with aldehyde fuchsin for lipofuscin pigment and with Darrow red for basophilic material). This micrograph shows the noradrenergic and neuromelanin-laden projection cells of the coeruleus nucleus (*coer*) as well as the serotonergic and lipofuscin-laden projection neurons of the dorsal (*rd*) and central (*rc*) raphe nuclei. **b–d** The lesions in the locus coeruleus worsen and ultimately result in pronounced neuronal loss, as shown here between stages 2 and 4. *Arrows* in **c** point to Lewy neurites. **e** Centers influenced by descending fiber tracts generated from the nuclei of the gain (or level) setting system. The lower raphe nuclei, gigantocellular nucleus of the reticular formation, the coeruleus-subcoeruleus complex modulate excitability levels of spinal and medullary centers that influence somatosensory and viscerosensory input as well as those influencing visceromotor and somatomotor output (*Ncl. grac./cun.*). **Abbreviations:** *Ncl. grac./cun.*, cuneate and gracile nuclei; *trigem. nuclei*, trigeminal sensory nuclei. *Scale bar* in **b** is valid for **c** and **d**

nuclei, all of which are components of what is known as the "gain" or "level setting" system of the medulla oblongata (Fig. 14e; Holstege 1992, 1996; Holstege et al. 2004; Nieuwenhuys 1996; Braak et al. 2000; Koutcherov et al. 2004). The sensory relay nuclei, vestibular nuclei, precerebellar nuclei (e.g., inferior olivary nucleus, pontine gray), cerebellum, and somatomotor nuclei of the lower brainstem (cranial nerves V–VII, XII) develop, if at all, only slight changes. Major components of the cerebellar circuit are introduced and discussed briefly in Sect. 8.6 below (Fig. 29). The cerebellar cortex and centers of the cerebellar circuit commence myelination early (prenatally) with the result that their long-axoned projection neurons develop medium-caliber and thick-caliber myelin sheaths. Accordingly, they tend to withstand involvement altogether or the Lewy pathology appears there in stages 5–6 (see Sect. 8.6 and Fig. 28b,c).

5.2 Basic Organization of the Gain Setting System

5.2.1 Lower Raphe Nuclei

Owing to the massive development of the pontine gray in the human brain, the complex of raphe nuclei is separated into an upper group, with predominantly ascending projections, and a lower group, with descending projections. The lower nuclei include the pallid, obscure, and great raphe nuclei. The pallid raphe nucleus is a loosely structured midline structure close to the anteromedial fissure, its lateral portions extending into the narrow spaces between the medial lemniscus and the pyramidal tract. The obscure raphe nucleus is more densely populated and consists of two narrow, perpendicularly arranged cellular lamellae that are embedded within the medial lemniscus. The great raphe nucleus is the largest of the three nuclei and extends caudally from the level of the facial motor nucleus (VII; Fig. 13a) and superior olive, slightly exceeding the level of the pontomedullary junction where it reaches its greatest extent as a spindle-shaped nucleus with two bilaterally symmetrical portions (rm; Fig. 13b). Predominant among the neuronal types forming the raphe nuclei are medium-sized multipolar projection cells—most probably serotoninergic, although nonserotonergic cells also exist (Ohm et al. 1989)—that contain large amounts of coarse lipofuscin granules (rm; Fig. 13a,b). Lewy pathology evolves only within the lipofuscin pigment-laden (i.e., serotonergic) neurons of the lower raphe complex. In comparison with controls, the number of lipofuscin pigment-laden raphe neurons appears to be reduced in individuals with advanced sPD.

5.2.2 Magnocellular Reticular Nuclei

The gigantocellular reticular nucleus is the preeminent component among the magnocellular nuclei of the reticular formation (gig; Fig. 13a). Its oral portion extends into the caudal third of the pontine tegmentum, whereas its caudal extremity blends imperceptibly into the central nucleus of the medulla. The nucleus is bounded ventromedially by the medial lemniscus and the dorsal accessory olive. Dorsolaterally, it abuts on the intermediate reticular zone and the parvocellular reticular nucleus. Two types of large multipolar neurons and a heterogeneous population of small nerve cells make up the nucleus. The predominant type among the large neurons contains large deposits of lipofuscin pigment granules that often extend into the cone-shaped stem of the dendrites. Intensely stained clusters of Nissl substance occupy the peripheral portion of the cell body. The second, less frequently occurring, type among the large cells bears some resemblance to a motor neuron and probably is representative of the premotor and motor neurons that populate the lower brainstem. The multipolar soma displays numerous distinct, intensely basophilic Nissl bodies. These nerve cells either contain a few dust-like lipofuscin granules or remain entirely devoid of pigment. The group of small nerve cells comprises many different types of neurons based on the varying pigmentation patterns and the nature of the basophilic material. The large pigment-laden nerve cells of the gigantocellular reticular nucleus bear the brunt of the Lewy pathology that develops within the reticular formation (Fig. 13c–e). Both the sparsely pigmented premotor and motor neurons as well as the heterogeneous group of small neurons are spared in sPD (Braak et al. 2000; E. Braak et al. 2001).

5.2.3 Coeruleus–Subcoeruleus Complex

This noradrenergic nucleus (equivalent to the rodent A6 and A7 groups) is located in the pontine tegmentum, close to the lateral angle of the fourth ventricle (Fig. 14a). From its ill-defined anterior pole commencing at approximately the level of the decussation of the IVth cranial nerve, the column-like coeruleus nucleus (A6 group) extends in a caudal direction to the level of the VIIth cranial nerve. The subcoeruleus portion (A7 group) appears less compressed and occupies more space beneath the anterior half of the coeruleus nucleus. The chief cell type of the nuclear complex is a catecholamine-synthesizing neuron that contains neuromelanin and has a medium-sized to large, rounded cell body with a marginally placed nucleus and coarse Nissl bodies at the periphery. A pale central region of the soma appears almost to lack Nissl material or pigment granules. These characteristics make the neurons readily distinguishable from the nerve cell types of surrounding nuclei. The melanized neurons of the coeruleus–subcoeruleus complex generate long, thin, and sparsely myelinated axons that diffusely project to the striatum, cerebellum, all areas of the cortex, and portions of the lower brainstem and spinal cord (Halliday 2004).

The pathological process in the coeruleus–subcoeruleus complex chiefly affects the neuromelanin-containing neurons (Zweig et al. 1993). LBs and LNs do not occur in the smaller nerve cells residing within the complex that display features different from those of the catecholaminergic neurons. In sPD cases, the neuromelanin granules usually are tinier and less densely packed than those in controls. Abnormal, particulate α-synuclein aggregations develop within the accumulations of neuromelanin. Comparison with control cases reveals an increasing loss of melanized neurons in later stages of the disorder (Fig. 14b–d). The observation that all of the gain setting system nuclei, and only these, become affected almost simultaneously in stage 2 is consistent with the conjecture that retrograde axonal and transneuronal transport via preexisting pathways plays an important role in the pathogenesis of sPD.

5.3 Potential Functional Consequences of the Lesions

The lower brainstem is positioned between the higher order centers of the tel-, di-, and mesencephalon, on the one hand, and the spinal cord on the other. The majority of its nuclei promote control and modification of data exchanged between the supramedullary and inframedullary components of the CNS. The potential functional consequences of the Lewy pathology in stages 1–2 are implicit in Fig. 14e: The lower row depicts the spinal and medullary centers of ascending sensory and descending motor pathways. Each of these relay nuclei is regulated by the supervening nuclei of the gain setting system (upper row in Fig. 14e), which receive major input from higher order components of the limbic and motor systems, such as the central subnucleus of the amygdala (Figs. 12b and 22) and the pedunculopontine tegmental nucleus (Figs. 27b and 28a). The level setting system is equipped to limit the conduction of incoming pain signals in stress situations and places the organism's motor neurons in a heightened state of preparedness for action (Randich and Gebhart 1992; Holstege et al. 2004). The descending tracts of the gain setting nuclei form a sensory control system that partially inhibits or entirely blocks the relay nuclei for somatosensory and viscerosensory input. In addition, these nuclei serve as a motor control system for both the somatomotor and visceromotor output. They regulate the sensitivity as well as excitability levels of medullary and spinal premotor and motor neurons that receive input from the neocortex, the magnocellular portion of the red nucleus, and other supramedullary sources (Figs. 14e and 29).

The lesions that accumulate within the gain setting system can be anticipated to reduce the functional capacities of the nuclei involved (Figs. 13c-d and 14b-d). Normally, the target neurons of these nuclei are capable of switching back and forth from a lower to a higher level of excitability, and the individual's emotional status determines the appropriate level in any given situation. An intact limbic system is necessary for the maintenance of this adaptive behavior, and patients suffering from sPD characteristically exhibit a disassociation between the voluntary and emotional motor systems (Holstege et al. 2004). Currently, much attention is being focused on reversing the impaired somatomotor functions that result from the destruction of nigral dopaminergic nerve cells. However, transplantation of fetal midbrain cells into the putamen of levodopa-refractory patients is fraught with the risk of developing new lesions (Kordower et al. 2008; Li et al. 2008) and cannot alleviate the neuronal dysfunction, nerve cell loss, or abnormal motor neuron responsiveness in other parts of the nervous system as a whole, including the limbic system-moderated gain setting nuclei in the lower brainstem (Zarow et al. 2003; Lang and Obeso 2004).

6
Stage 3

6.1
Involvement of the Mesencephalic Tegmentum and Basal Forebrain

In stage 3 the disease process crosses the upper limit of the pontine tegmentum and makes inroads into the mesencephalic tegmentum and basal forebrain (Fig. 5a). The Lewy pathology in previously involved sites worsens, and inclusion bodies appear in the catecholaminergic neurons of both the dorsal vagal area (A2 group) and intermediate reticular zone (A1 group), as well as in nerve cells close to the roof of the fourth ventricle that form the cerebellar portion of the coeruleus nucleus (A4 group). LNs and LBs also develop in the central subnucleus of the amygdala (Fig. 15a,c), posterolateral subnucleus of the substantia nigra (pars compacta; Fig. 16c–g), and in the pedunculopontine tegmental nucleus (Fig. 17a–d), upper raphe nuclei (Fig. 17e,f), hypothalamic tuberomamillary nucleus (Fig. 19d,e), and magnocellular nuclei of the basal forebrain (Figs. 15a,c and 16a,b).

At this point, the entire tectum and thalamus are uninvolved (see Sect. 7.1). Slight changes are visible in the somatomotor nuclei of the eye muscles (cranial nerves III, IV, VI). Phylogenetically older components of the limbic system, such as the habenular nuclei, interpeduncular nucleus, and dorsal tegmental nucleus of Gudden (asterisk in Fig. 14a) as well as extremely phylogenetically recent diencephalic structures, e.g., the hypothalamic lateral tuberal nucleus (see Sect. 6.2 and Fig. 19d, e) remain intact for the duration of the disease. Equally striking is that both the diencephalic and mesencephalic components of the striatal and cerebellar circuits that generate moderately to thickly myelinated axons (pallidum, subthalamic nucleus, red nucleus) are also resistant.

6.1.1
Central Subnucleus of the Amygdala

A hallmark of stage 3 is the involvement of the central subnucleus of the amygdala (Fig. 15c), which is recognizable even without the use of a light microscope in unconventionally thick (100 μm) polyethylene glycol-embedded hemisphere sections processed for α-synuclein immunoreactivity (Fig. 15a). A tight-meshed network of thin LNs intermingled with LBs appears in the central subnucleus, features that make it possible to distinguish it from the medial subnucleus of the amygdala as well as the surrounding intercalated cell masses that do not become involved in sPD. In the same hemisphere sections, it is also evident that cortical areas have not been drawn into the pathological process (Fig. 15a). The central subnucleus receives projections from the basolateral nuclear complex of the amygdala as well as from magnocellular nuclei of the basal forebrain and the temporal mesocortex (Braak et al. 1994). The subnucleus generates dense descending, sparsely myelinated projections directed to both the gain setting nuclei and the dorsal motor nucleus of the

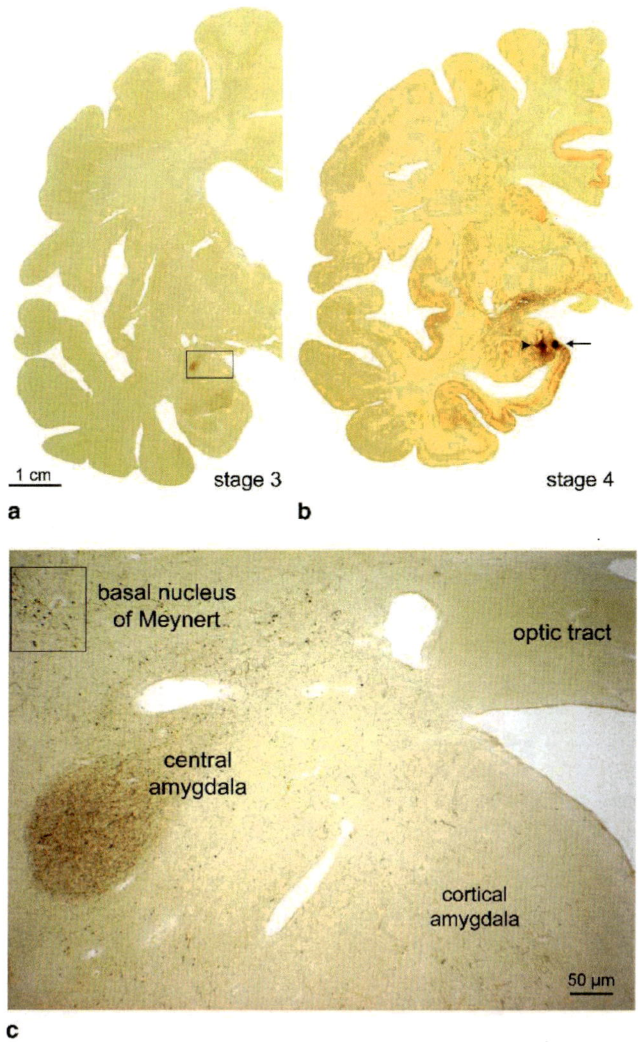

Fig. 15 a–c Lewy pathology in 100 µm hemisphere polyethylene glycol-embedded sections at stage 3 and 4 in immunoreactions against α-synuclein. **a** The cerebral cortex is still uninvolved at stage 3. The darkened area within the frame indicates the macroscopically visible central subnucleus of the amygdala in an asymptomatic individual. **b** Additional subnuclei of the amygdala become affected in neuropathological stage 4 (intense immunostaining, arrow indicates the accessory cortical subnucleus, arrowhead points to the accessory basal subnucleus of the amygdala). The Lewy pathology in the amygdala is accompanied by the first pathological changes in the cortex, seen here in the entorhinal region, anteromedial temporal, insular, and cingulate mesocortex. Note the faintly immunoreactive interstitial nucleus of the diagonal band and basal nucleus of Meynert directly above the amygdala. **c** This micrograph is an enlargement of the framed area in **a**. The pathology within the central subnucleus is not only severe and characteristic for the amygdala at stage 3 but illustrates the selective vulnerability of this subnucleus in contrast to immediately surrounding regions. Framed area in **c** appears again in Fig. 16a. *Scale bar* in **a** also applies to **b** as well as Fig. 23a, b

vagal nerve. In so doing, it exerts superordinate limbic influence on each of these modulatory nuclei in the lower brainstem. The central subnucleus also regulates all of the nonthalamic nuclei with diffuse cortical and subcortical projections. As such, it influences nearly all centers belonging to the central autonomic network (Fig. 27b; Amaral et al. 1987; Price et al. 1987; Sims and Williams 1990; Loewy 1991; Bohus et al. 1996; Fudge and Haber 2000; Liubashina et al. 2000).

6.1.2 Substantia Nigra and Adjoining Nuclei

The substantia nigra is located in the inferior tegmentum where it borders on the cerebral peduncles and extends from the posterior tip of the mamillary bodies to the level of the oculomotor nucleus (van Domburg and ten Donkelaar 1991). The nucleus consists of three zones: a cell-dense pars compacta (equivalent to the rodent A9 cell group), a cell-sparse pars diffusa, and a reticulate portion. The pars compacta can be subdivided further into seven subnuclei (Fig. 16c): an anterior group of three (anteromedial, anterointermediate, and anterolateral subnucleus) lies close to the superior border of the substantia nigra, whereas the posterior group occupies a deeper position and consists of two cell plates positioned at a blunt angle (pm; posteromedial and pl; posterolateral subnucleus; Fig. 16c), a hook-like cellular formation (ps; posterosuperior subnucleus; Fig. 16c), and a wedge-shaped cellular assembly (m; magnocellular subnucleus, Fig. 16c) (Braak and Braak 1986).

Approximately 450,000 dopaminergic melanized neurons populate the nigral pars compacta on each side of the brainstem (Ma et al. 1999; Cabello et al. 2002; Rudow et al. 2008). Isolated nerve cells of this type also occur in the pars diffusa

Fig. 16 **a, b** Lewy pathology in the basal nucleus of Meynert (100 μm thick polyethylene glycol-embedded, immunoreaction against α-synuclein). Note the increasing degree of involvement between stages 3 (**a**) and 5 (**b**). Here, **a** corresponds to the frame seen in Fig. 15c. **c** Substantia nigra from a control case in a 400 μm-thick polyethylene glycol-embedded section stained with aldehyde fuchsin for lipofuscin pigment and with Darrow red for basophilic material. This overview displays important cell groups of the pars compacta, i.e., the posteromedial (*pm*), posterolateral (*pl*), posterosuperior (*ps*), magnocellular (μ), and perirubral subnuclei (*ru*). Medially, follow the paranigral nucleus (*pn*) and pigmented parabrachial nucleus (*pp*). Midline structures are the interpeduncular nucleus (*ip*) and the linear raphe nucleus (*rl*), a lamellar-like component of the upper raphe nuclei, that extends within the decussation of the superior cerebllar peduncle (*dscp*). **d–g** Progression of Lewy pathology within the substantia nigra, pars compacta in 100 μm sections immunostained for α-synuclein. **d, e** Overview (**d**) and detail (**e**) of the posterolateral subnucleus at stage 3 in a nonsymptomatic individual. The micrographs do not yet display loss of melanized neurons (dark spots) but clearly reveal the presence of Lewy neurites. A large number of long and spindle-shaped Lewy neurites appear between the melanized neurons accompanied by a few cell bodies. **f, g** At stage 4, the loss of large numbers of melanized neurons is obvious (**f**), and, in stage 5, the pars compacta is almost denuded of melanin-containing nerve cells. Note that the Lewy pathology has also "disappeared" from the tissue. Scale bars: *a* is valid for **b**, **g** is valid for **e, f**

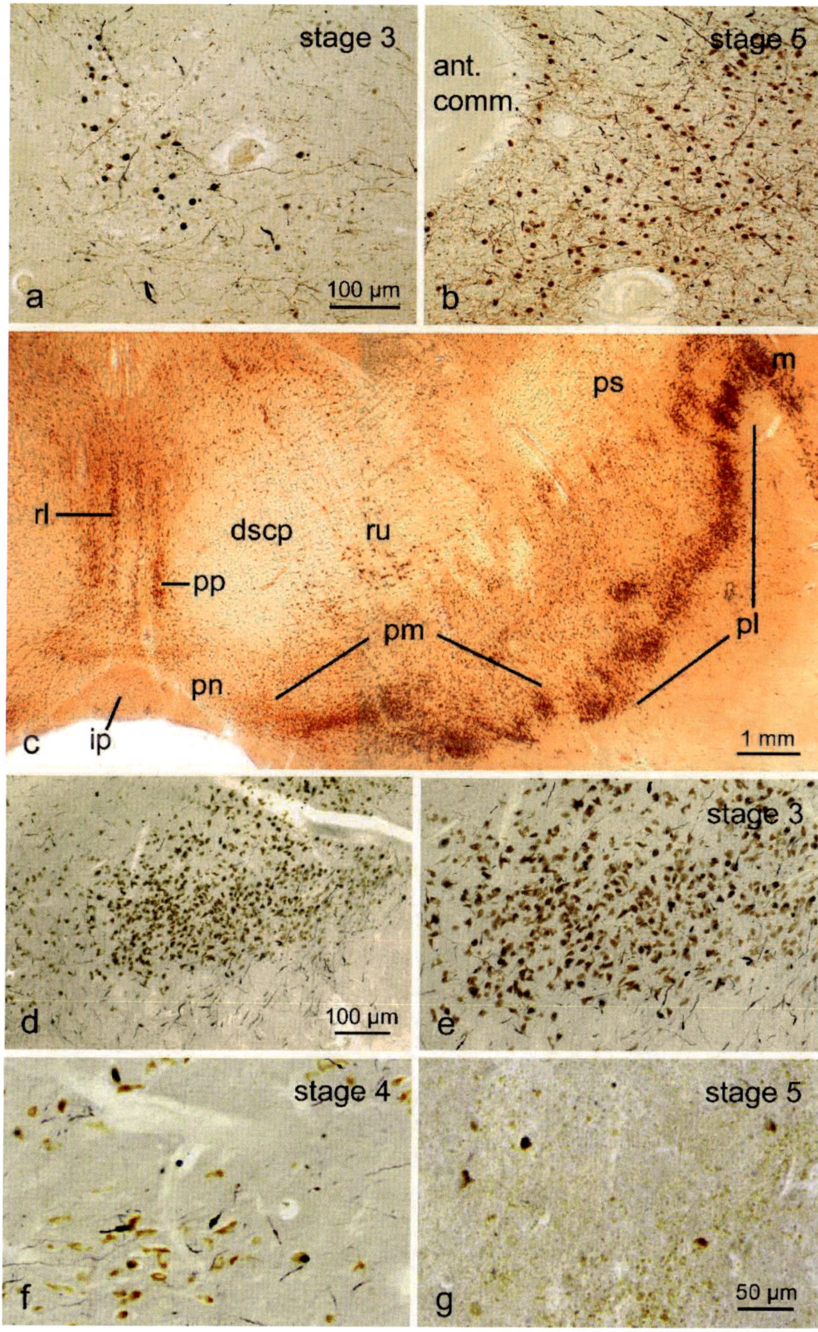
stage 3
100 µm
a
ant.
comm.
stage 5
b
ps
m
rl
dscp
ru
pp
pm
pl
pn
ip
c
1 mm
100 µm
d
stage 3
e
stage 4
f
stage 5
50 µm
g

and pars reticulata. The neurons in the pars compacta have a spherical, marginally located cell nucleus, sharply outlined peripherally located Nissl bodies, and, at one cell pole, a large deposit of neuromelanin granules. Their somata give off a few thick dendrites arranged in bundles that extend into the other two zones but usually not beyond the boundaries of the substantia nigra itself. The neurons of the pars compacta generate fine-caliber, thinly myelinated axons with thick cone-shaped initial segments. All three nigral zones receive heavy input from both the ventral and dorsal striatum. Even projections that terminate within the pars reticulata mainly contact the dendrites of pars compacta neurons. Caudatonigral fibers probably terminate chiefly within the anterior group of subnuclei, whereas putaminonigral fibers reach the posterior group. Projections from the accumbens nucleus appear to be weighted toward the anteromedial subnucleus. The nigral dopaminergic neurons generate a predominantly ipsilateral and topographically organized projection, the comb bundle, to the striatum. The anteromedial subnucleus mainly projects to the ventral striatum, the anterolateral subnucleus to the head of the caudate nucleus, and the posteromedial subnucleus targets the putamen.

The reticulate portion mainly contains the dendrites of pars compacta neurons (Francois et al. 1987; Yelnik et al. 1987). Scattered among this network are relatively few medium-sized GABAergic projection neurons that lack neuromelanin and project to thalamic nuclei (VA/VLa), the pedunculopontine tegmental nucleus, superior colliculus, and parvocellular reticular formation of the brainstem. Nigrotectal fibers terminate chiefly on tectospinal projection neurons and play a critical role in the control of visual saccades (Wurtz and Hikosaka 1986).

In stage 3, a few, and in part very long, LNs appear in the posterolateral subnucleus followed by LBs in the dopaminergic melanized neurons (Fig. 16d,e). Lesions then follow in the posterosuperior and posteromedial cellular groups, whereas the Lewy pathology skirts the magnocellular and anterior subnuclei—at least for the time being (Damier et al. 1999a,b). The nonmelanized projection neurons in the pars reticulata and pars diffusa remain unaffected. Macroscopically, the substantia nigra appears intact and shows no clear-cut signs of neuronal loss (Fig. 16d). Thinning and loss of melanoneurons set in at stage 4 and worsen thereafter (Fig. 16f,g; Braak et al. 2003a, 2004).

Apparently, a mild loss of midbrain dopaminergic neurons accompanies aging (Kubis et al. 2000; Rudow et al. 2008). Thus, the enormous neuronal destruction that occurs in the substantia nigra during sPD has rightly been ascribed to the aggregation of α-synuclein in this especially vulnerable nerve cell population. Once the pathological process reaches the substantia nigra, a cascade of events may be set in motion that boosts the immediate effects of the pathological process and, in so doing, compels involved nigral cells to decline more rapidly than nerve cell types located elsewhere. Idiosyncratic features (e.g., high iron content, neuromelanin, dopamine turnover) and local conditions (proteasomal dysfunction, free radical formation, inflammation) are among the causative factors that could contribute to portions of the nigra's increased susceptibility to the pathological process (Zecca et al. 2006; Lang 2007). Moreover, individual differences obviously exist; in some individuals the nigral destruction is fulminant and massive, whereas in others it is prolonged and milder. There is undoubtedly a whole series of additional factors that determine and influence the timing and extent of the neurodegeneration in the substantia nigra. The latest estimate is a loss of one-third of striatal dopaminergic terminals (Hilker et al. 2005)—a substantially lower figure than the 50%–80% loss of midbrain dopaminergic neurons traditionally cited as the prerequisite for the appearance of clinically recognizable motor symptoms. Loss of nigral melanized neurons correlates significantly with the severity of hypokinesia and rigidity (Greffard et al. 2006).

Groups of melanized neurons in the midbrain are also encountered close to the red nucleus (ru; perirubral subnucleus, equivalent to the rodent A8 group; Fig. 16c) and in the ventromedial tegmentum (pn; paranigral nucleus and pp; pigmented parabrachial nucleus, both equivalent to the rodent A10 group; Fig. 16c). The paranigral nucleus forms an arch covering the interpeduncular nucleus (ip; Fig. 16c) and continues into the leaf-like, sagittally oriented pigmented parabrachial nucleus that accompanies the linear nucleus of the upper raphe system (rl; Fig. 16c). Little is known about the afferent connections of the paranigral and pigmented parabrachial nuclei, which provide the major dopaminergic projections to the amygdala, hippocampal formation, and entorhinal region. Furthermore, neocortical motor areas, anterior cingulate fields, and prefrontal association areas receive dense dopaminergic projections from the paranigral nucleus. This nucleus and the pigmented parabrachial nucleus display less pronounced changes than the substantia nigra while the perirubral subnucleus and other dopaminergic neurons of the mesencephalic central gray display little or no pathology (Hirsch et al. 1988; Gibb 1991; Agid et al. 1993; McRitchie et al. 1997; Damier et al. 1999a,b). There is currently no plausible explanation for the resistance on the part of these dopaminergic neurons.

6.2 Basic Organization of Nonthalamic Nuclei with Diffuse Projections

Although spaced widely apart from each other, the nonthalamic nuclei are unified in that they generate long, thin, and sparsely myelinated diffuse fiber tracts that project toward many subcortical nuclei as well as nearly the entire cerebellar and cerebral cortex. These nuclei include the noradrenergic coeruleus–subcoeruleus complex, the serotonergic upper raphe nuclei, the dopaminergic paranigral and parabrachial pigmented nuclei of the mesencephalic tegmentum, the histaminergic hypothalamic tuberomamillary nucleus, and the cholinergic magnocellular nuclei of the basal forebrain (Fallon and Loughlin 1987; Saper 1987, 1990; Mesulam et al. 1992; Nieuwenhuys 1999). In spite of the different transmitter substances, each of these nuclei is highly susceptible to the pathological process. With the exception of the coeruleus–subcoeruleus complex, Lewy pathology in all of these regions begins to appear in stage 3 (Braak et al. 2003a, 2004, 2006c).

6.2.1 Magnocellular Nuclei of the Basal Forebrain

Clusters of large multipolar neurons located in the basal forebrain form three major nuclei: the medial septal nucleus, the interstitial nucleus of the diagonal band, and the basal nucleus of Meynert. The small-celled substantia innominata partially surrounds these nuclei. Leaf-like extensions of the magnocellular nuclear complex impinge deeply on the external and internal medullary layers of the pallidum, thereby forming the peripallidal subnucleus. The magnocellular forebrain nuclei contain a variety of neuronal types. The predominant type is a large cholinergic nerve cell that generates a lengthy axon and reaches virtually all subdivisions of the cerebral cortex. A second type is GABAergic and contains various neuropeptides. The magnocellular nuclei receive abundant information from olfactory, insular, subgenual, and prefrontal cortical areas, entorhinal region and amygdala, ventral striatum, ventral pallidum, and the gain setting nuclei. They project diffusely to the cortex and numerous subcortical sites, including the mediodorsal thalamic nuclei (transferring limbic data from ventral pallidum to prefrontal cortex), the reticular nucleus of the thalamus, and the amygdala. The nuclei can be viewed as a relay between the components of the autonomic system, limbic circuit, and neocortex (Mesulam 2004; Heimer and van Hoesen 2006).

The involvement of the amygdala and substantia nigra in stage 3 is supplemented by the initial involvement of the basal forebrain nuclei (Fig. 16a). At first, serpentine or spindle-shaped LNs predominate over LBs and, remarkably, nearly the entire lengths of the axons generated from these nuclei are α-synuclein-immunopositive from the very outset, so that axon bundles can be observed over wide stretches (Fig. 16a). All of the nuclei become involved simultaneously, and the Lewy pathology there resembles that seen in the peripallidal portion, where α-synuclein-immunoreactive neurons can be distinguished from the large striatal interneurons that are generally spared. In contrast to the large neurons of the basal forebrain, the small neurons of the substantia innominata remain intact.

6.2.2 Pedunculopontine Tegmental Nucleus

The pedunculopontine nucleus occupies dorsolateral portions of the mesencephalic tegmentum. (Fig. 17a,b). The nucleus is composed of two groups of neurons, one that synthesizes acetylcholine and another that contains noncholinergic neurotransmitters (GABA, glutamate) (Mesulam et al. 1989). The internal pallidum provides the main input (Fig. 28a). The pallidal fibers skirt the cholinergic neurons and terminate in noncholinergic neurons of the pedunculopontine nucleus. Efferent fiber tracts of the nucleus project with thin and sparingly myelinated axons to nigral dopaminergic neurons, the subthalamic nucleus, pallidum, striatum, intralaminar nuclei of the thalamus, and nonthalamic nuclei with extensive diffuse subcortical and cortical projections (Fig. 28a). These ascending projections outnumber by far the descending ones that go out to the gain setting nuclei and spinal cord (Nakano 2000; Pahapill and Lozano 2000; Aravamuthan et al. 2007). Owing to its strategic position between limbic and striatal circuits, the pedunculopontine nucleus influences both cognitive processes and locomotion. Along with the gain setting nuclei, it also functions as part of a larger rhythmogenic complex that induces and modulates patterns of oscillatory activity, including the sleep–waking cycle and states of arousal, consciousness, attention, and vigilance (Garcia-Rill 1991; Steckler et al. 1994; Inglis and Winn 1995; Rye 1997; Winn et al. 1997; Lee et al. 2000; Nakano 2000).

LNs and LBs begin to appear in the pedunculopontine nucleus during stage 3. At first the LNs seen there are remarkably long. These become voluminous and, at higher stages, form a dense network (Fig. 17d). Gradually, LBs appear in the somata of the cholinergic nerve cells, particularly within the compact portion of the pedunculopontine tegmental nucleus (Hirsch et al. 1987; Jellinger 1988; Zweig et al. 1989).

6.2.3 Tuberomamillary Nucleus

The tuberomamillary nucleus extends through the posterior tuberal and anterior mamillary territories of the hypothalamus and is closely associated with the median forebrain bundle. The nucleus almost entirely covers and surrounds the cellular islands of the lateral tuberal nucleus (Fig. 19d). The density of the neurons in the tuberomamillary nucleus diminishes toward its outermost portions, and these ill-defined areas gradually overlap with nerve cells belonging to the hypothalamic gray matter. The large tuberomamillary neurons are richly supplied with lipofuscin granules and have a ragged outline. Strongly staining Nissl material is concentrated at peripheral portions of the soma and in invaginations of the nuclear envelope. Many of the tuberomamillary nerve cells contain acidophilic granules within their somata, a feature also encountered within the morphologically related cholinergic neurons of the magnocellular basal forebrain nuclei (Issidorides et al. 1991). The neurons of the tuberomamillary nucleus are mainly GABAergic and histaminergic but also synthesize additional transmitters, including adenosine and galanin (Panula et al. 1990; Passani et al. 2000). Its neurons generate far-flung projections, including those to the striatum, and the cortical projection is of the same magnitude as that of the magnocellular nuclei of the basal forebrain (Saper 1990).

Lewy pathology in the tuberomamillary nucleus (Langston and Forno 1978; Kremer and Bots 1993) appears from stage 3 onwards. Most review articles dealing with the neuropathology of sPD do not mention the severe involvement of the tuberomamillary nucleus (Fig. 19 d,e), and studies devoted to the potential clinicopathological correlations of the lesions there are still needed.

6.2.4 Upper Raphe Nuclei

The upper raphe complex comprises a dorsal, central, and linear nucleus. Occupying a territory between the medial longitudinal fascicles and the ependymal lining of the fourth ventricle, the dorsal raphe nucleus extends almost as far caudally as the coeruleus–subcoeruleus complex (rd; Fig. 14a) and abuts, rostrally, a cap-like formation on nuclei of the oculomotor nerve. Its supratrochlear subnucleus is very expansive in the human brain and forms a voluminous cover over the motor nucleus of the trochlear nerve (Fig. 17e,f). The central raphe nucleus is situated ventrally from the medial longitudinal fascicles (rc; Fig. 14a). Symmetrical portions on both sides are separated by means of a narrow fiber lamella that is devoid of cells. The linear raphe nucleus lies within the decussation of the superior cerebellar peduncle (dscp, rl; Fig. 16c). The melanized neurons of the pigmented parabrachial nucleus (pp; Fig. 16c) build a sheath surrounding the rostral portion of the linear raphe nucleus. The predominant cell type in the upper raphe complex is a medium-sized, multipolar projection neuron that contains large quantities of lipofuscin pigment granules. Intermingled with this neuronal type is a small quantity of nerve cells of similar size and shape but sparsely pigmented or nearly lacking lipofuscin granules. Most of the nerve cells in the upper raphe nuclei are serotonergic, but additional transmitters have also been reported within the complex. The upper raphe nuclei project diffusely not only to cortical targets but also to all subdivisions of the striatum. Lewy pathology appears in these nuclei from stage 3 onwards directly following the involvement of the lower raphe nuclei at stage 2 (Fig. 17e,f).

6.3 Potential Functional Consequences of the Lesions

Whereas in stage 2 the gain setting nuclei are at the center of the disease process (see Sects. 5.2 and 5.3), a bifurcation of the pathological process appears to take place in stage 3; one arm inflicts damage on the superordinate centers of the somatomotor system (i.e., the pedunculopontine tegmental nucleus), the other impairs the relay nuclei of the central autonomic network (i.e., the central subnucleus of the

---

Fig. 17 a–f Pedunculopontine tegmental and dorsal raphe nuclei in sporadic Parkinson's disease. **a, b** Overview and detail showing the topographical location of the pedunculopontine tegmental nucleus (*ped. pont.*) lateral to the decussation of the superior cerebellar peduncle, and situated between the locus coeruleus (posteriorly) and the substantia nigra (anteriorly). Control case, 400 μm-thick polyethylene glycol-embedded section stained with aldehyde fuchsin for lipofuscin pigment and with Darrow red for basophilic material. **c, d** Overview (**c**) and detail (area in **d** corresponds to the *asterisk* in **c**) micrographs showing the dense network of Lewy neurites that develops within the pedunculopontine nucleus beginning in stage 3 and increasing in density at higher stages (here, stage 5). **e,f** Overview (**e**) and detail (**f**) of the Lewy pathology in the upper raphe nuclei from stage 3 and upwards, here, the supratrochlear portion (*supratrochl.*) of the dorsal raphe nucleus at stage 6. *Asterisks* in **c** and **e** indicate the pedunculopontine tegmental nucleus. **c–f** 100 μm polyethylene glycol-embedded sections immunostained for α-synuclein

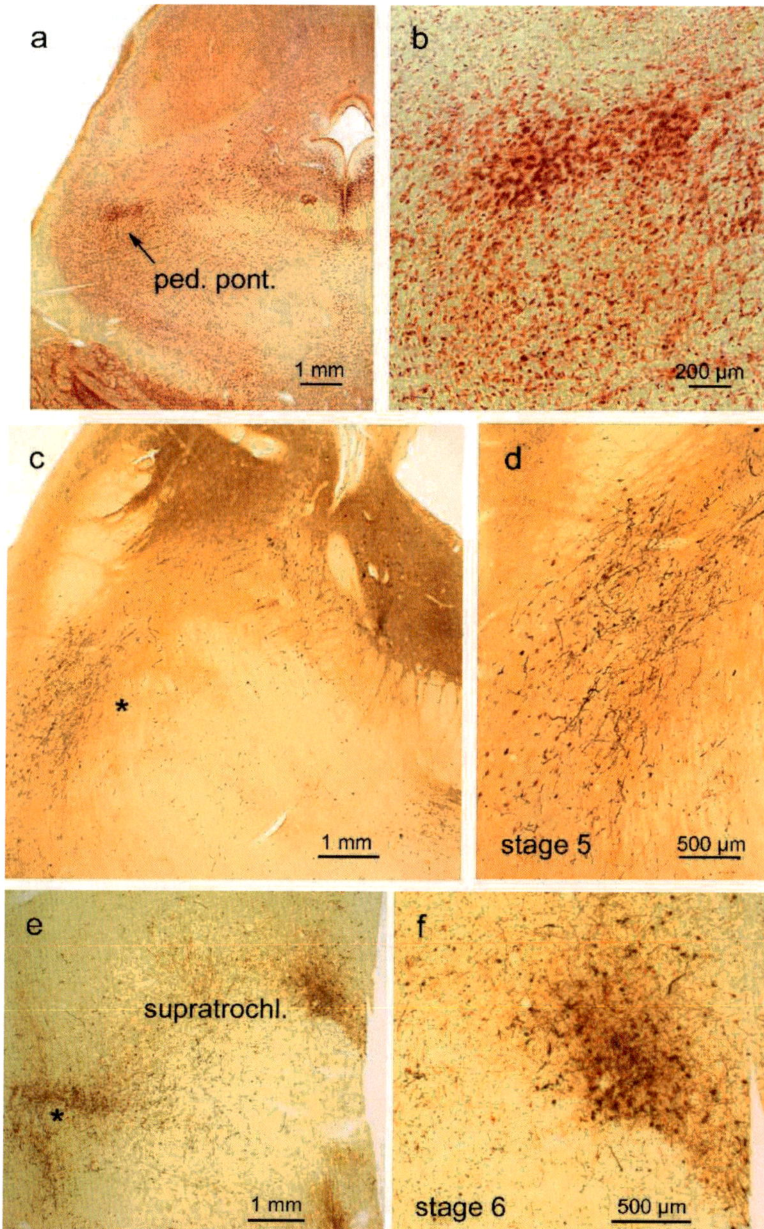
a
ped. pont.
1 mm
b
200 µm
c
*
1 mm
d
stage 5
500 µm
e
supratrochl.
*
1 mm
f
stage 6
500 µm

amygdala). Both foci of the Lewy pathology control the gain setting nuclei while maintaining connections to the substantia nigra and exerting influence on all of the diffusely projecting nonthalamic nuclei (Fig. 27b). In other words, at the same time that the central subnucleus of the amygdala channels the pathological process into high order components of the limbic and visceromotor systems, the pedunculopontine nucleus routes it toward those of the somatomotor system.

Involvement of the noradrenergic coeruleus–subcoeruleus complex occurs from stage 2 onwards and results in loss of melanized projection neurons (Fig. 14c,d) and an accompanying diminution of noradrenergic input to the cerebral cortex. The dopaminergic system follows in stage 3. Melanized neurons in the medial substantia nigra and ventral tegmental area are the source of major projections to the temporal and frontal cortex, and loss of these cells is responsible for cortical dopamine deficiency (Gaspar et al. 1991). The cholinergic magnocellular nuclei of the basal forebrain and the pedunculopontine tegmental nucleus also become involved in stage 3, thereby causing decreased cholinergic input to the cerebral cortex and to specific subcortical sites. During the same stage, Lewy pathology develops in additional diffusely projecting nuclei, the upper raphe complex and the hypothalamic tuberomamillary nucleus. These nuclei influence cortical processing and modulate the activity level of cortical projection neurons in accordance with external and/or internal conditions and input. The cumulative damage inflicted on all of these sites exacts its toll on the general input to the cortex and leads to curtailment of the versatility with which cortical functions adapt to ever-changing demands. Such damage can pave the way for a reduction of higher cognitive functions (Rinne et al. 1989; Duyckaerts et al. 1993; Pillon et al. 1994; Perry et al. 1996; Dubois and Pillon 1997; Wolters and Francot 1999; Grudzien et al. 2007).

By exploiting existing anatomical pathways, a neurotropic pathogen that enters the brain via the dorsal motor nucleus of the vagal nerve could pass through the gain setting nuclei to reach the central subnucleus of the amygdala and the pedunculopontine tegmental nucleus. From each of these footholds, it could advance into the pars compacta of the substantia nigra and to all of the nonthalamic nuclei that have diffuse subcortical and cortical projections (Figs. 27b and 29). Viruses inoculated into the gastric mucosa follow precisely this route to reach the central subnucleus of the amygdala (Rinaman et al. 2000).

7
Stage 4

Since clinical protocols of some individuals with stage 3 Lewy pathology and those of many stage 4 cases mention incipient disease-related motor symptoms, we postulated in 2002 that perhaps at some point during stages 3–4 the nonsymptomatic phase of "incidental" sPD yields to the clinically recognizable phase of the illness (Fig. 4e; Braak et al. 2003a; Braak and Del Tredici 2008).

7.1 Involvement of the Amygdala and Thalamus

The lesions in previously involved nuclei become more severe in stage 4 (Fig. 22). The loss of neuromelanin-containing neurons in the coeruleus–subcoeruleus complex becomes macroscopically visible. Cases without LNs/LBs in the ventral tegmental area and upper raphe nuclei in the preceding stage begin to display them at stage 4. Impairment of the diffusely projecting nonthalamic nuclei probably begins to reduce the input to the cerebral cortex. Of equal importance in stage 4, however, is the involvement of additional subnuclei of the amygdala (Fig. 15b), the intralaminar and midline nuclei of the thalamus (Fig. 19a), and a specific portion of the cerebral cortex. In most cases, the vulnerable cortical site is the anteromedial temporal mesocortex (Fig. 20a). At about the same time, but with somewhat greater irregularity, subtle changes also appear in insular, subgenual, and anterior cingulate cortical areas, as well as in the second sector of the Ammon's horn (Fig. 20b,c). Initial changes may also be visible in the striatum, thalamic relay nuclei with projections to the cerebral cortex, and the claustrum. The involvement of these sites is described more extensively below under stages 5 and 6 (Sects. 8.3–8.5). Lewy pathology is not seen in neocortical areas, and the same applies to the pallidum and subthalamic nucleus. Similarly, the small melanized neurons (corresponding to the rodent A12 group) of the infundibular and periventricular nuclei that comprise the dopaminergic system of the hypothalamus remain untouched by the pathological process (Matzuk and Saper 1985). It is noteworthy that these nerve cells generate a short axon and do not establish connections to other nuclei that consistently develop LNs and LBs.

7.1.1 Amygdala

The amygdala is situated deep in the temporal lobe, directly anterior to the uncal portion of the hippocampus, and consists of mediocentral nuclei, cortical portions, and the basolateral complex. The human basolateral complex is especially spacious and consists of the lateral, basal, and accessory basal nuclei (Fig. 18a). The interstitial nucleus of the terminal stria can be regarded as an elongation of the amygdala and, together with the central subnucleus, it makes up the extended amygdala (Heimer and van Hoesen 2006; Alheid 2003). The terminal stria bidirectionally connects amygdalar nuclei with the hypothalamus (Saper 2004). The neuronal components of the amygdala resemble cortical nerve cells. Morphologically speaking, the projection neurons and local circuit neurons at both sites are comparable, but the longitudinal axis of the pyramidal-like amygdalar projection neurons is not aligned as uniformly as its counterpart in the cortex. The amygdala receives a wide range of afferents that serve to integrate exteroceptive information (e.g., olfactory, somatosensory, auditory, and, in particular, visual data) with interoceptive stimuli from various autonomic centers. The nuclei of the basolateral complex maintain reciprocal connections with the hippocampal formation and neocortical high-order association areas. The lateral subnucleus receives input from the sensory neocortex either directly or via the anteromedial temporal mesocortex and entorhinal region. Dense projections generated in the basal and accessory basal subnuclei terminate in the ventral striatum, ventral pallidum, mediodorsal thalamus, insular, and, above all, prefrontal cortex (Amaral et al. 1987; Aggleton 2000; Harding et al. 2002; Heimer and van Hoesen 2006).

In stage 4, Lewy body pathology develops in the accessory cortical nucleus (arrow in Fig. 15b), basolateral nuclear complex, and interstitial nucleus of the terminal stria. From the outset, the basal and accessory basal (arrowhead in Fig. 15b) subnuclei display higher densities of LBs and a thinner network of LNs than the lateral subnucleus, as becomes more evident in the following stage (Fig. 18b). The cortical nuclei and the parvocellular portions of the basal and accessory basal nuclei remain less involved. The lesional patterns, as well as the severity of the pathology, that develop in the accessory cortical nucleus and interstitial nucleus of the terminal stria resemble those in the central subnucleus (Braak et al. 1994).

7.1.2 Thalamus

The thalamus makes up four-fifths of the human diencephalon. An external medullary lamina covers the lateral surface and is separated from the internal capsule by a narrow sheet of nerve cells that form the thalamic reticular nucleus. An internal medullary lamina divides the thalamus into lateral, medial, and anterior relay nuclei and includes the group of anterior intralaminar nuclei (Fig. 19a). The thalamic centromediano-parafascicular complex mediates between anterior and posterior intralaminar nuclei (Fig. 26d). The limitans nucleus of the posterior intralaminar group faces the pretectum of the midbrain (Fig. 19c). The group of thalamic midline nuclei consists of leaf-shaped cellular assemblies that are spread out between the epithelial lining of the third ventricle and the mediodorsal nuclei of the thalamus. Taken together, the anterior intralaminar, posterior intralaminar, and centromediano-parafascicular groups constitute the thalamic intralaminar nuclei.

All of the thalamic relay nuclei are reciprocally connected with the cerebral cortex and display, in essence, the same composition. Large multipolar glutamatergic projection neurons with long aspiny dendrites that radiate in all directions predominate. Their axons are heavily myelinated and send small, columnar terminal arborizations into layers II–V of well-defined cortical areas. Interspersed among the glutamatergic projection cells are smaller bipolar local circuit neurons (Braak and Braak 1984).

The intralaminar nuclei receive major afferents from the tegmental pedunculopontine nucleus and, in turn, generate a powerful projection to the dorsal striatum. The centromediano-parafascicular complex receives afferents from, among other sources, the internal pallidum, limbic areas, and the cerebellum. Whereas projections from the centromedian nucleus target the putamen, those of the parafascicular

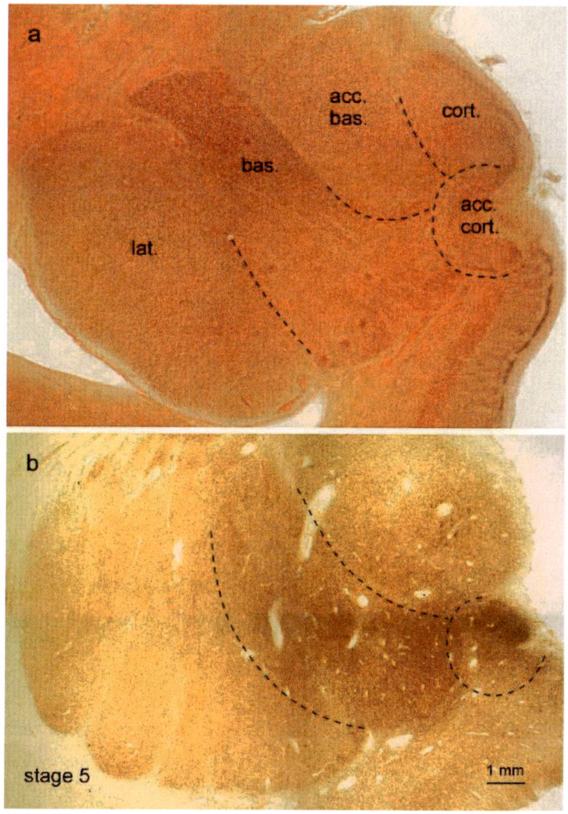

Fig. 18 a,b Amygdala in sporadic Parkinson's disease. **a** Overview of the basolateral complex as seen in a 100 μm-thick polyethylene glycol-embedded section stained with aldehyde fuchsin for lipofuscin pigment and with Darrow red for basophilic material. Borders of the various subnuclei, i.e., the lateral (*lat.*), basal (*bas.*), and accessory basal (*acc. bas.*) subnuclei, are relatively clearly discernible and can be differentiated from the cortical (*cort.*) and accessory cortical (*acc. cort.*) subnuclei. **b** Overview of α-synuclein aggregations in these portions of the amygdala of a case with stage 5 brain pathology. Note the severe involvement of the accessory cortical subnucleus. Involvement of this subnucleus commences as early as stage 3 together with the appearance of Lewy pathology in the central subnucleus. 100 μm polyethylene glycol-embedded section immunostained for α-synuclein. *Scale bar* in **b** is also valid for **a**

nucleus terminate at the caudate nucleus (Parent and Hazrati 1995a). All of the midline and most of the intralaminar nuclei harbor glutamatergic multipolar projection neurons with thin, spiny dendrites and poorly myelinated axons. Their projections are less specific than those of the relay nuclei, i.e., they innervate more than one cortical area and terminate more diffusely in layers I and VI. Activation of these projections plays a role in attentiveness to visual, auditory, and somatosensory stimuli (Parent and Hazrati 1995a). Projections from the midline nuclei furnish thalamo-allocortical circuits and pathways to the ventral striatum (Berendse and Groenewegen 1991; Groenewegen and Berendse 1994; van der Werf et al. 2002).

The reticular nucleus partially covers the lateral portions of the thalamus and is ideally situated to receive collaterals from both thalamocortical and corticothalamic fibers. In addition, the reticular nucleus receives dense cholinergic afferents from the magnocellular nuclei of the basal forebrain and pedunculopontine tegmental nucleus (Fig. 28a). It is also reached by a strong GABAergic projection from the external segment of the pallidum (Fig. 28a). By contrast, the GABAergic neurons of the reticular nucleus project only to other nuclei of the thalamus and not to the cortex. This situation certainly indicates that some of the modulating influence on the basal ganglia may occur at the level of the reticular thalamic nucleus, the latter serving as a gating device or an "attentional gate" that enables the individual to differentiate between essential and inessential data arriving from the periphery (Hazrati and Parent 1991; McAlonan and Brown 2002).

The Lewy pathology that develops in the thalamus at stage 4 does so only in select nuclei. The midline nuclei and the anterior and posterior group of the intralaminar nuclei are the most vulnerable (Fig. 19a,c) and all of their projection neurons are furnished with poorly myelinated axons (Henderson et al. 2000; Rüb et al. 2002). The sheer mass of the inclusion body pathology that accumulates in these nuclei during sPD contrasts sharply with the very mild pathology in most of the thalamic relay nuclei (Fig. 19a). Physiologically, the centromedian nucleus, which customarily is also assigned to the intralaminar nuclei, is nearly devoid of α-synuclein, and it is probably for that reason that no intraneuronal aggregates develop there (Fig. 26d; Braak et al. 2001b). Whereas only isolated LNs are occasionally visible in the thalamic relay nuclei in stage 4 cases, the reticular nucleus appears to be somewhat more severely involved. α-Synuclein immunoreactive dot-like inclusions in thalamic relay neurons and astrocytes are described in greater detail below (see Sect. 8.3).

7.2 Basic Organization of the Cerebral Cortex and Limbic Circuit

7.2.1 Allocortex and Neocortex

Considerable differences exist between the two basic types of cerebral cortex: the expansive areas of the neocortex versus the small and heterogeneously composed allocortex (Fig. 20a,c). The neocortex is chiefly responsible for processing data related to the world beyond the individual. It receives somatosensory, visual, and auditory input while simultaneously overseeing somatomotor activity that impinges on the organism's environment (Fig. 21a). The human neocortex takes up nearly 95% of the total cortical surface area in the brain and, apart from a

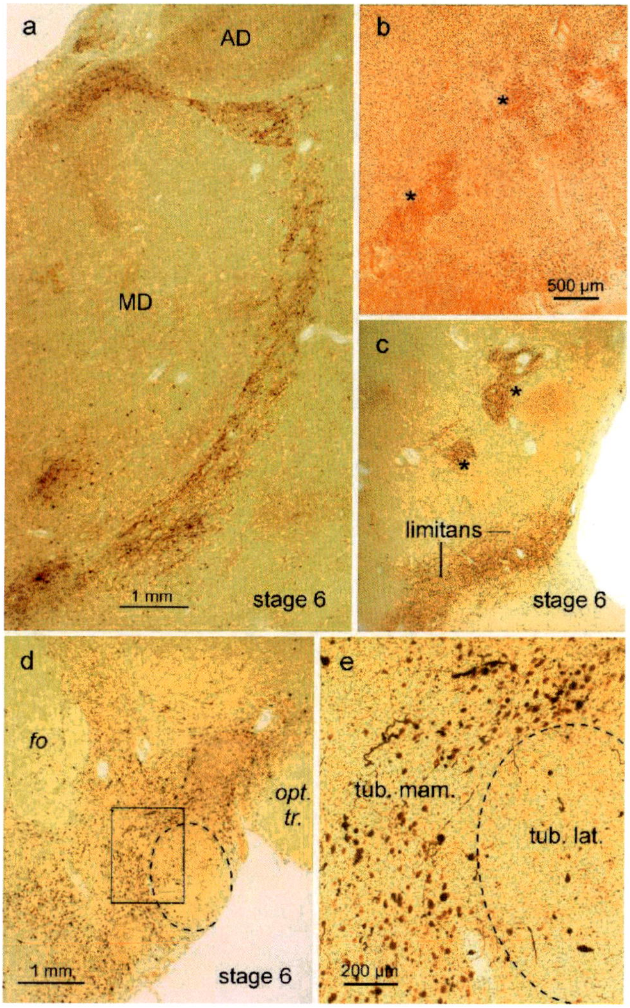

Fig. 19 a–e Intralaminar nuclei of the thalamus and the hypothalamic tuberomamillary nucleus in sporadic Parkinson's disease. **a, c** Lewy pathology at stage 6 in 100 µm polyethylene glycol-embedded hemisphere sections immunostained for α-synuclein, supplemented by an overview of a 100 µm-thick lipofuscin pigment-Nissl section from a control individual in **b**. **a** Anterior group of intralaminar nuclei. The severity of the Lewy pathology in the intralaminar nuclei contrasts sharply with the mild involvement of the mediodorsal (*MD*) and anterodorsal (*AD*) nuclei. **b, c** *Asterisks* indicate the position of the posterior intralaminar nuclei in the pulvinar. The limitans nucleus (**c**) located close to the border toward the midbrain is particularly heavily involved. **d, e** Overview (**d**) and detail (**e**) of the lateral hypothalamus in an immunoreaction against α-synuclein, 100 µm-thick polyethylene glycol-embedded section at stage 6. The severely involved tuberomamillary nucleus (*tub. mam.*) partially caps the rounded cellular groups of the lateral tuberal nucleus (*tub. lat.*). Remarkably, the latter nucleus remains nearly unaffected. **Abbreviations:** *fo*, fornix; *opt. tr.*, optic tract. *Scale bar* in **b** is also valid for **c**

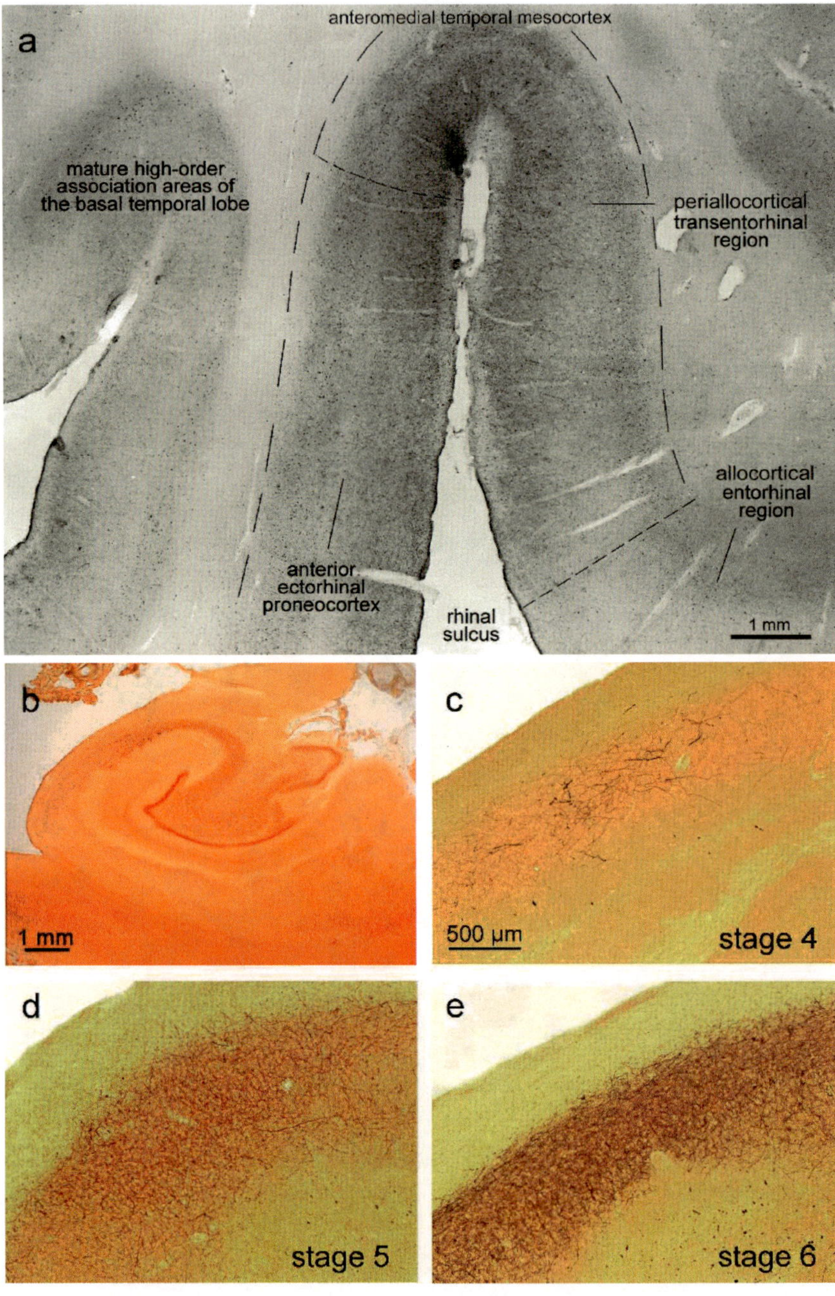
a
anteromedial temporal mesocortex
mature high-order association areas of the basal temporal lobe
periallocortical transentorhinal region
allocortical entorhinal region
anterior ectorhinal proneocortex
rhinal sulcus
1 mm
b
1 mm
c
500 µm
stage 4
d
stage 5
e
stage 6

minimum of regional variations, is organized in six layers. The mature neocortex of the parietal, occipital, and temporal lobes is divided into highly refined and well-myelinated primary fields that are responsible for the initial processing of incoming data from the sensory organs via particularly dense input from specific thalamocortical projections. A zone of somewhat less highly differentiated and only moderately myelinated unimodal sensory first order association areas flank each of the primary fields. These, in turn, are interconnected with extensive but relatively simply organized and incompletely myelinated unimodal and/or heteromodal high order sensory association areas (Fig. 21b). The frontal lobe is similarly structured into a primary motor field and premotor areas, the latter of which give way to high-order association areas, i.e., the prefrontal cortex (Fig. 21b). The enormous dimensions of the prefrontal cortex, which is the equivalent of the highest executive instance in the CNS, is a distinct feature of the human brain (Braak 1980; Pandya and Yeterian 1985, 1996; Mesulam 1998; Zilles 2004).

Allocortical areas range from trilaminate fields to those with more than six layers. The mammalian brain contains two allocortical centers: (1) the olfactory bulb together with related olfactory areas, and (2) the hippocampal formation as well as the entorhinal and presubicular regions (Braak and Braak 1992; Braak et al. 1996). The nuclear complex of the amygdala is closely interconnected with the allocortical regions. Whereas the olfactory bulb mainly receives exteroceptive data, the hippocampal formation lacks direct sensory input and has to rely, instead, on indirect input from olfactory areas and neocortical association areas to obtain information about the world beyond the individual. The superordinate centers of the limbic system also receive data from the internal organs and can influence endocrine and visceromotor functions (Fig. 21a). The olfactory bulb and related structures in the human brain are rudimentary in comparison to the corresponding areas of macrosmatic mammals. By contrast, the hippocampal formation and related regions are well developed. The hippocampus comprises three major units: the dentate fascia, Ammon's horn (CA), and subiculum. A transverse tissue section, cut at the level of the lateral geniculate body, permits recognition of all three units and of the four

Fig. 20 Lewy pathology in the anteromedial temporal mesocortex and Ammon's horn. **a** The anteromedial temporal mesocortex is the point of entry for the cortical pathology that develops in sporadic Parkinson's disease and begins to appear in stage 4 (here seen in a 100 μm hemisphe re section immunostained for α-synuclein at stage 5). The overview includes part of the allocortical entorhinal region, followed by the periallocortical transentorhinal region, and proneocortical ectorhinal region (both belonging to the mesocortex), and even shows part of the temporal mature neocortex (*upper left*). Cortical Lewy pathology usually begins in the transentorhinal region. **b** Overview of the Ammon's horn in a 100 μm polyethylene glycol-embedded lipofuscin pigment-Nissl section. See also Fig. 25e. **c–e** Development of the plexus of Lewy neurites in the second sector of the Ammon's horn (CA2) beginning in stage 4 (**c**), continuing in stage 5 (**d**), and reaching a very high density in stage 6 (**e**). 100 μm polyethylene glycol-embedded hemisphere sections immunostained for α-synuclein. *Scale bar* in **c** also applies to **d** and **e**

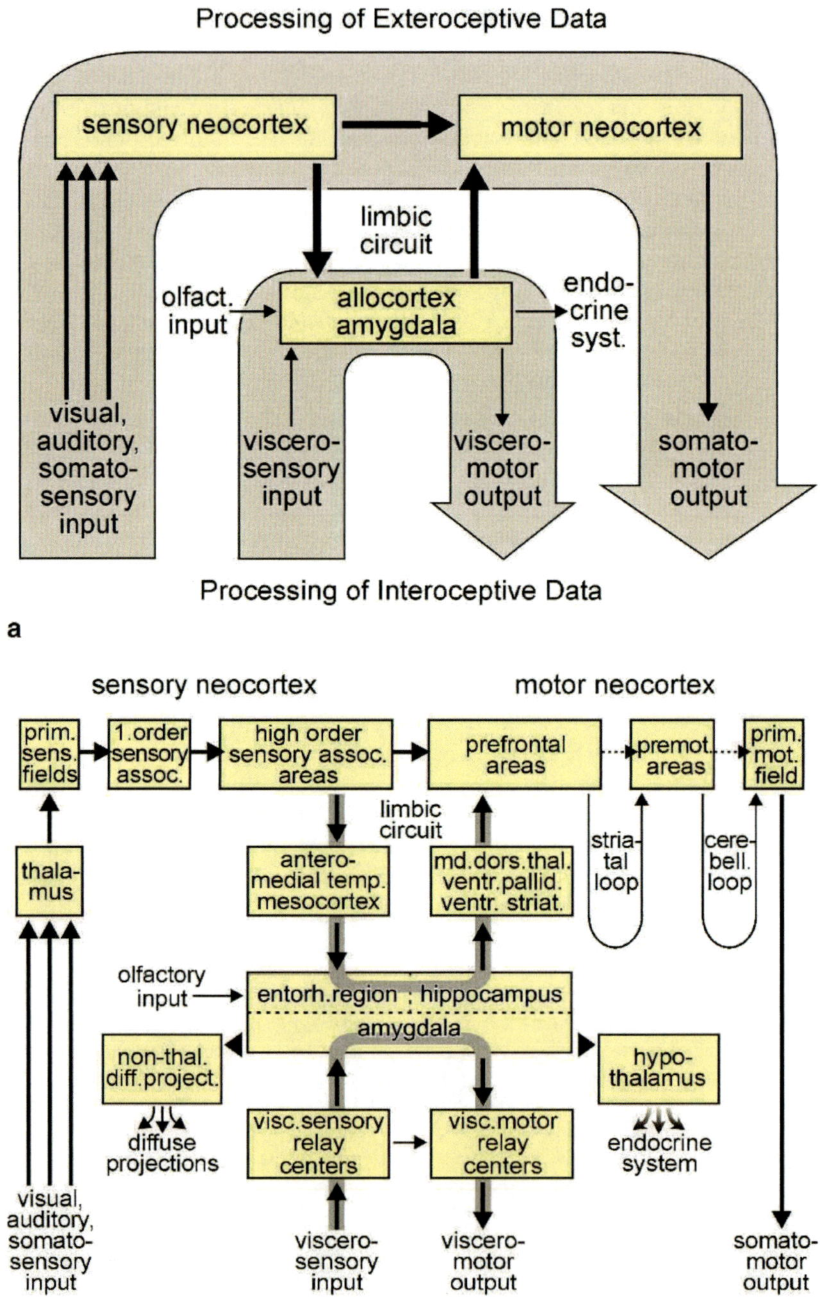
Processing of Exteroceptive Data
sensory neocortex
motor neocortex
limbic circuit
olfact. input
allocortex amygdala
endo-crine syst.
visual, auditory, somato-sensory input
viscero-sensory input
viscero-motor output
somato-motor output
Processing of Interoceptive Data
a
sensory neocortex
motor neocortex
prim. sens. fields
1.order sensory assoc.
high order sensory assoc. areas
prefrontal areas
premot. areas
prim. mot. field
limbic circuit
thala-mus
antero-medial temp. mesocortex
md.dors.thal. ventr.pallid. ventr. striat.
stria-tal loop
cere-bell. loop
olfactory input
entorh.region ; hippocampus
amygdala
non-thal. diff.project.
hypo-thalamus
diffuse projections
visc.sensory relay centers
visc.motor relay centers
endocrine system
visual, auditory, somato-sensory input
viscero-sensory input
viscero-motor output
somato-motor output
b

sectors of the Ammon's horn (Fig. 20b). The second sector (CA2) stands out owing to its densely packed and voluminous pyramidal cells. Little is known about its afferents and efferents or their functional significance. Some of the afferent fibers in the CA2 probably originate from the hypothalamic tuberomamillary nucleus (Veazey et al. 1982). The sector receives minimal input from mossy fibers of the dentate fascia or Schaffer collaterals from CA3 (Sanides 1969; Braak 1980; Braak et al. 1996; Gloor 1997; Insausti and Amaral 2004).

7.2.2 Mesocortex

Transitional zones between the neocortex and allocortex comprise a unique architectonic entity—the mesocortex—which includes a periallocortical region allied with the allocortex proper and a proneocortical region that leads to the mature neocortex (Fig. 20b). Since allocortical layers tend to transgress the limits of their parent territories, the periallocortical region contains an admixture of both allocortical and neocortical layers. The proneocortex is usually dysgranular in character, i.e., the parvocellular layers II and IV are poorly developed (Braak and Braak 1985). By contrast, the anteromedial temporal portion of the mesocortex is exceptionally well developed only among higher primates and, above all, humans. Allocortical and mesocortical areas form an unbroken ring that encircles the medial and basal components of the hemispheres (Fig. 20a). Directly adjacent to the olfactory system are the periallocortical and proneocortical portions of the

Fig. 21 **a** Schematic drawing of the cerebral cortex. The neocortex is chiefly responsible for processing and influencing exteroceptive data. It receives visual, auditory, and somatosensory input and regulates somatomotor output. The afferent and efferent trunks of the limbic circuit interconnect the neocortex, allocortex, and amygdala. In addition to olfactory input (*olfact. input*), the allocortex and amygdala mainly receive and process interoceptive data. They also influence both the endocrine system and centers regulating visceromotor output. **b** Somewhat more detailed rendition of the scheme shown in **a**, this time displaying major subdivisions of the neocortex and major components of the limbic circuit (entorhinal region, hippocampal formation, amygdala). The afferent trunk of the limbic circuit includes the anteromedial temporal mesocortex, which is exceptionally well developed in the human brain. The efferent trunk chiefly includes the ventral striatum, ventral pallidum, and mediodorsal thalamus, which direct incoming data toward the prefrontal areas of the neocortex. See also Fig. 22 and the accompanying legend. **Abbreviations:** *1. order sensory assoc.*, first order sensory association areas; *cerebell. loop*, cerebellar loop; *entorh. region*, entorhinal region; *high order sensory assoc. areas*, high order sensory association areas; *md. dors. thal.*, mediodorsal nuclei of the thalamus; *non-thal. diff. project.*, nonthalamic diffusely projecting nuclei; *premot. areas*, premotor areas; *prim. mot. field*, primary motor field; *prim. sens. fields*, primary sensory fields; *ventr. pallid.*, ventral pallidum; *ventr. striat.*, ventral striatum; *visc. motor relay centers*, visceromotor relay centers of the brainstem; *visc. sensory relay centers*, visceroseonsory relay centers of the brainstem

insular and subgenual mesocortex. Both represent the brain's preeminent organizational level for processing interoceptive data from the internal organs and for regulating visceromotor and endocrine functions. A gradual increase in cortical differentiation is observable if an imaginary line is drawn from the periallocortex through the proneocortex, the high-order and the first order association areas, to the most highly advanced primary fields of the neocortex. This architectonic hierarchy is reflected by an equally hierarchical arrangement of the major pathways that interconnect the basic territories (Sanides 1969; Braak 1980; Pandya and Yeterian 1996; Mesulam 1998).

7.2.3 Limbic Circuit

Black arrows in Fig. 21a indicate the manner in which exteroceptive input is relayed from the most highly differentiated neocortical primary sensory fields to both the primary motor field and allocortex in the limbic circuit. Somatosensory, visual, and auditory data flow upstream from each of the primary sensory fields, via short projections, to reach layer IV of the adjoining sensory first order association areas and, from there, to succeeding sensory high order association areas (Rockland and Pandya 1979). These areas, in turn, send relatively long and sparsely myelinated projections to the prefrontal cortex (Fig. 21b), where the upstream connections terminate in layer IV of their target fields. Minor pathways that lead away from the prefrontal cortex are provided by short corticocortical downstream projections ending in layer I of their target areas (broken arrows in Fig. 21b; Barbas and Rempel-Clower 1997). These pathways transmit the data via the premotor areas to the primary motor field, which acts as a gateway for motor programs being relayed to medullary and spinal premotor and motor neurons. The chief routes for the downstream dataflow, however, are the striatal and cerebellar loops (Fig. 21b). Both circuits integrate the basal ganglia, thalamus, many of the lower brainstem nuclei, and the cerebellum into the regulation of cortical output (Alexander et al. 1990; Heimer et al. 1991; Albin et al. 1995; Parent and Hazrati 1995a; Alheid 2003; Alheid et al. 1990; Haber and Gdowski 2004; Petrides and Pandya 2004).

Superordinate limbic system centers (entorhinal region, hippocampus, amygdala) also are involved in this upstream data transfer and do so at the nodal point where exteroceptive information is conveyed from sensory high order association areas to the prefrontal cortex (Fig. 21b). As if through a defile, some of the data in transit from neocortical sensory high-order association areas proceed, after having left the mainstream, through multiple neocortical relay stations and are channeled through the anteromedial temporal mesocortex to the entorhinal region, hippocampal formation, and points beyond: the ventral striatum, ventral pallidum, mediodorsal thalamus, and finally the prefrontal cortex (Fig. 21b). During the evolutionary transition from macrosmatic mammals to microsmatic higher primates, the human neocortex not only undergoes a remarkable degree of expansion but also a thorough internal reorganization of its interconnectivities with centers of the limbic circuit. Of central significance is the enormous increase in size and sophistication of those portions of limbic circuit centers that receive input from and generate output to the neocortex. These organizational changes take place at the expense of the formerly predominant territories involved in the processing of olfactory information. As such, neocortical—and not olfactory—data are the dominant sources of input to the human limbic circuit (see the afferent trunk of the limbic circuit in Fig. 21b). The output structures of limbic circuit centers (e.g., subiculum of the hippocampal formation and basolateral nuclei of the amygdala) generate dense projections to the accumbens nucleus and ventral striatum. From there, the limbic data are conducted via the ventral pallidum and magnocellular mediodorsal thalamic nuclei to medial and orbitofrontal portions of the prefrontal neocortex—the executive instance of the human cortex (see the efferent trunk of the limbic circuit in Fig. 21b; Del Tredici and Braak 2004; Suzuki and Amaral 2004; Heimer and van Hoesen 2006; Groenewegen and Trimble 2007).

In addition to exteroceptive data, the centers of the limbic circuit receive crucial interoceptive input from the internal organs (Fig. 21b). As a result, they are optimally positioned to extract from the entire spectrum of incoming exteroceptive/interoceptive data the information needed to produce an appropriate response to any given situation or set of circumstances. This also means that the limbic centers influence virtually all of the subcortical centers that regulate or control endocrine and autonomic functions. Within the larger framework of the human limbic system, the limbic circuit is integral to the maintenance of learning, memory, and emotional equilibrium. At the same time, however, it affects somatomotor activity, inasmuch as its influence on the prefrontal cortex causes an individual's motor behavior to reflect his or her emotional state of mind. In their role as custodians of memory and learning, the centers of the limbic circuit act as a neuronal bridge that links the external and internal worlds (Heimer et al. 1982, 1991; Hyman et al. 1990; Zola-Morgan and Squire 1993; Haber et al. 1995; Mesulam 1998; Squire and Schacter 2002; Heimer and van Hoesen 2006).

7.3 Involvement of the Temporal Mesocortex and Hippocampal Formation

7.3.1 Temporal Mesocortex

In stage 4 the pathological process finally gains access to the cerebral cortex, where it usually begins in anteromedial portions of the temporal mesocortex (Figs. 5a, 20a, 22, and 23a). All of the cell layers in this cortical transition zone develop LNs/LBs, albeit in different densities. The density of the Lewy pathology is lowest in layer IV. Pyramidal cells that contain LBs are seen chiefly in layers V–VI and only occasionally in layer IIIc (Fig. 23c). Neither granular α-synuclein nor pale bodies are observable in the somata of vulnerable pyramidal cells (Wakabayashi et al. 1995). Inclusion bodies that otherwise display all of the morphological features of

LBs but lie free in the neuropil occur frequently. The nerve cells in the white matter below the temporal mesocortex (*lamina cellularis profunda*) also show signs of involvement but do not readily attract the viewer's attention because their densities are appreciably lower than those in layers V and VI.

7.3.2 Hippocampal Formation

Lewy pathology develops in one of the deep layers of the entorhinal region and in the second sector (CA2) of the hippocampal formation directly after the appearance of LNs and LBs in the anteromedial temporal mesocortex: long and repeatedly arborizing LNs in the hippocampal stratum oriens push deeply into the pyramidal layer (Fig. 20c). In subsequent stages, the LN-plexus is visible macroscopically and progresses into adjoining portions of CA1 and CA3. This plexus is so prominent in sPD during stages 5 and 6 that, even without the standard histological sections through the substantia nigra, the neuropathologist can make the diagnosis based solely on its presence (Fig. 20d,e). Because the immunocytochemistry profile for the LNs in CA2 is negative for tyrosine hydroxylase, they probably are not derived from axonal arborizations of brainstem catecholaminergic neurons (Dickson et al. 1994). The CA3 sector and the hilus of the dentate fascia (CA4) contain LNs, rarely LBs. The dentate fascia proper and the subiculum remain nearly intact. Projection neurons in the pyramidal layers of CA1 display LBs and short granular LNs. In the molecular layer of CA1 and close to the hippocampal fissure, large spindle-shaped LNs lie dispersed in the neuropil. A further Lewy pathology predilection site is the transitional region between the amygdala and hippocampal formation.

7.4 Involvement of Insular, Subgenual, and Anterior Cingulate Areas

From the anteromedial temporal mesocortex, the disease process encroaches upon more distant, related cortical regions and extends into the insular, subgenual, and anterior cingulate areas (Fig. 24a,b; Vogt et al. 1992, 1993). The involvement of the dysgranular insular fields is less severe than that of the agranular portions. Although large, for the most part spherical, LBs are abundantly present throughout the insular, subgenual, and anterior cingulate areas, those in layers V and VI are the most prevalent. Very slender LNs run at right angles to the cortical surface and outnumber thick-caliber forms. Some appear to be attached to LBs. Club-shaped or spiral LNs also occur sporadically in all layers (Fig. 24b; Del Tredici and Braak 2004).

The agranular and dysgranular regions of the insula together with the adjoining association areas encompass gustatory areas and a topically organized representation of the internal organs and inner surface of the body. These regions are reciprocally connected with the subgenual and anterior cingulate areas, the entorhinal region, amygdala, claustrum, and thalamic limitans nucleus, as well as the pigmented parabrachial nucleus. They also generate major projections to the magnocellular nuclei of the basal forebrain and ventral striatum. In this manner, a pathway is established between the insular fields via the ventral pallidum and mediodorsal thalamus to the prefrontal association cortex. By way of the claustrum and magnocellular nuclei of the basal forebrain, the insular regions exert their influence on the cerebral cortex as a whole (Fig. 29). For this reason they have received the designation "viscerosensory and limbic integration cortex" (Witter et al. 1988; Oppenheimer et al. 1992; Mesulam and Mufson 1993; van Hoesen et al. 1993; Augustine 1996).

The subgenual region is part of the ventromedial frontal lobe and, like the insular regions above, represents a topically organized visceromotor center (Figs. 24a and 29). Reciprocally organized projections connect the region with adjoining prefrontal areas, the insular and anterior cingulate cortex, entorhinal region, hippocampal formation, amygdala, intralaminar and midline nuclei of the thalamus, lateral hypothalamus, periaqueductal nuclei, and autonomic regions of the lower brainstem and spinal cord. The subgenual region sends strong projections to the ventral striatum and, consequently, acts upon the prefrontal cortex via the ventral pallidum and, mediodorsal, thalamus (Fig. 29). As such, the subgenual region can be regarded as fulfilling the functions of a "visceromotor and limbic integration cortex" (Cechetto and Saper 1990; Morecraft et al. 1993; Price et al. 1996; Meredith et al. 1996; Chu et al. 1997; Öngür et al. 1998).

7.5 Potential Functional Consequences of the Lesions

Depending on the presence of additional factors, e.g., patient-specific neuronal reserves and preexisting comorbidities, the subcortical and cortical Lewy pathology that develops during stage 4 is probably sufficient to pave the way for some degree of clinically manifest cognitive impairment and, sometimes, visual hallucinations (Harding et al. 2002). Increasing deterioration of the nigrostriatal pathway not only can lead to greater impairment of somatomotor function but also can reduce the data transfer from limbic centers, via the dopamine-deprived ventral striatum, to the ventral pallidum, mediodorsal thalamus, and prefrontal neocortex (Fig. 22). At the same time, the direct dopaminergic innervation of the cerebral cortex also becomes hampered. In addition, the progressive involvement of the anteromedial temporal mesocortex can seriously impede the data flow originating in neocortical high-order sensory association areas and heading through the limbic circuit toward the prefrontal cortex. Severe damage to the temporal mesocortex in sPD can even deplete the frontal lobe of its limbic input altogether, thereby inducing cognitive decline, impaired executive functioning, or a demential syndrome (Hurtig et al. 2000; Mattila et al. 2000; Korczyn 2001; Ince and McKeith 2003; Braak et al. 2005; Korczyn and Reichmann 2006).

Since the monograph entitled *On the Shaking Palsy* by James Parkinson (1817), autonomic dysfunction has been consistently noted as an early and frequently occurring phenomenon (Rajput and Rozdilsky 1976; Goetz et al. 1986; Ludin et al. 1987; Korczyn 1990; Meco et al. 1991; Singer et al. 1992; Tanner et al. 1992; van Dijk et al. 1993; Martignoni et al. 1995; Quigley 1996; Pfeiffer 1998, 2003; Soykan et al. 1999; Chaudhuri 2001; Mathias 2002; Siddiqui et al. 2002; Kaufmann et al. 2004;

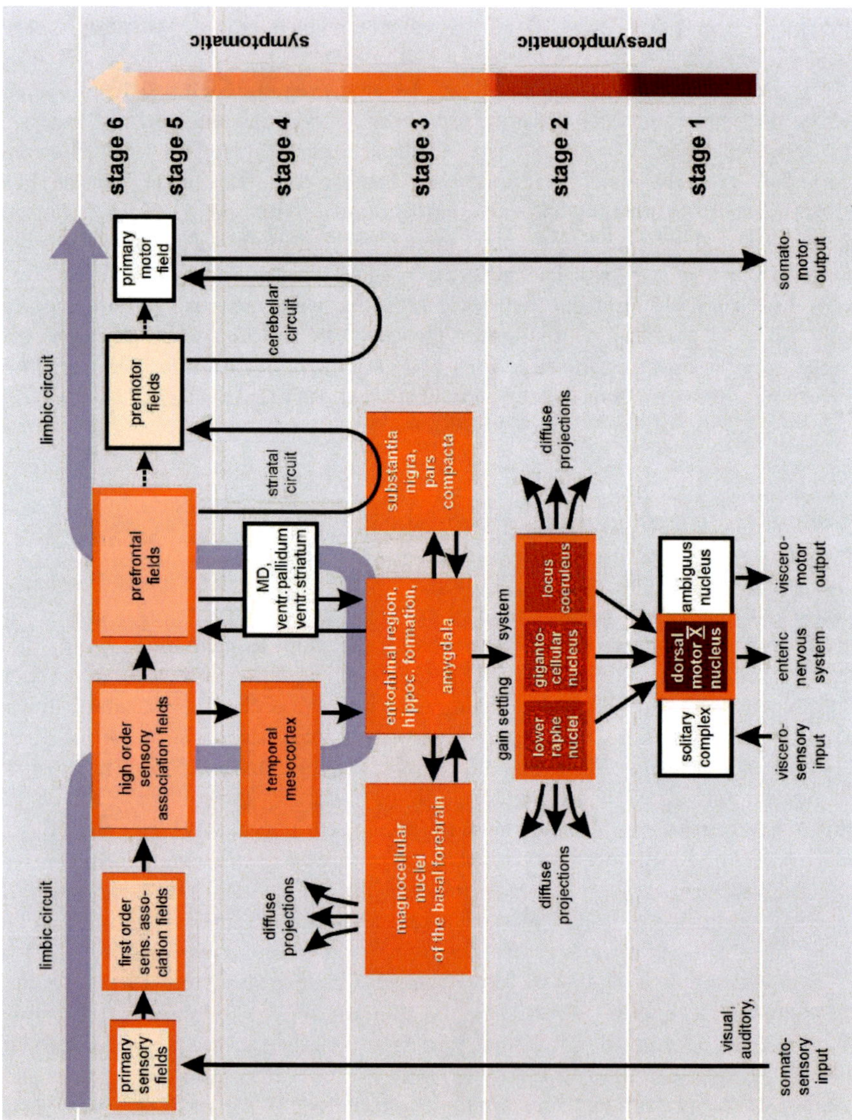
symptomatic
presymptomatic
stage 6
stage 5
stage 4
stage 3
stage 2
stage 1
limbic circuit
limbic circuit
primary sensory fields
first order sens. association fields
high order sensory association fields
prefrontal fields
premotor fields
primary motor field
temporal mesocortex
MD, ventr. pallidum ventr. striatum
striatal circuit
cerebellar circuit
diffuse projections
magnocellular nuclei of the basal forebrain
entorhinal region, hippoc. formation, amygdala
substantia nigra, pars compacta
gain setting system
diffuse projections
lower raphe nuclei
giganto-cellular nucleus
locus coeruleus
diffuse projections
solitary complex
dorsal motor X nucleus
ambiguus nucleus
visual, auditory,
somato-sensory input
viscero-sensory input
enteric nervous system
viscero-motor output
somato-motor output

Fig. 22 Diagram showing the major subdivisions of the neocortex and superordinate limbic circuit centers (entorhinal region, hippocampal formation, amygdala). The *shading intensity* in each of the regions indicates the increasing severity of the Lewy pathology in sporadic Parkinson's disease. Uninvolved areas appear in *white*. *Black arrows* indicate the major anatomical fiber tracts or interconnectivities and pathways. The limbic circuit occupies a strategic position between neocortical high-order association fields and prefrontal neocortex (*blue limbic circuit arrow*). The afferent trunk of the limbic circuit includes the temporal mesocortex, which is highly developed in humans. Normal limbic circuit functions are dependent on the structural integrity of the temporal mesocortex. The efferent trunk includes the ventral striatum (*ventr. striatum*), ventral pallidum (*ventr. pallidum*), and mediodorsal thalamus (*MD*), all of which transport data toward the neocortical prefrontal fields. Severe involvement of the temporal mesocortex results in reduced data transfer from the sensory neocortex via the entorhinal region, hippocampal formation (*hippoc. formation*), and amygdala to the prefrontal neocortex. A decline of cortically controlled cognitive functions can exacerbate other dysfunctions that develop in the course of sporadic Parkinson's disease owing to damage within both the visceromotor and somatomotor systems. **Abbreviations:** *dorsal motor X nucleus*, dorsal motor nucleus of the vagal nerve; *first order sens. association fields*, first order sensory association fields

Adler 2005; Magerkurth et al. 2005). Among the earliest complaints in sPD are obstipation and gastric dysmotility (Abbott et al. 2001, 2007; Pfeiffer 2003), urinary bladder dysfunction (Adler 2005; Winge and Fowler 2006), orthostatic hypotension (Saper 1998; Goldstein 2006; Martignoni et al. 2006), and cardiac sympathetic denervation (Goldstein et al. 2000; Li et al. 2002; Taki et al. 2004). Involvement of the final common pathway for the autonomic system, e.g., the preganglionic sympathetic and parasympathetic projection neurons of the brainstem and spinal cord, together with postganglionic nerve cells of the PNS and ENS in stage 2, is gradually compounded by deterioration of the gain setting nuclei in stage 2, as well as the amygdalar central subnucleus and interstitial nucleus of the terminal stria in stage 3. The Lewy pathology that develops in these nuclei may explain the early dysfunctions of the autonomic system. In stage 4 the lesions also systematically erode superordinate autonomic relay centers, including insular, subgenual, and anterior cingulate areas that influence respiration, heart rate, blood pressure, respiration, and gastrointestinal motility (Neafsy 1990; Loewy 1991; Reiman et al. 1997). Their structural and functional integrity are necessary for the maintenance of the sympathetically mediated increase in heart rate during physical exertion and emotional stress, the resting level for the gain of the cardiac component of the baroreflex response, and the appropriate sudomotor response of the skin to emotional stimuli (Damasio et al. 1990; Cechetto and Saper 1990; Oppenheimer et al. 1992; Drevets et al. 1997; Lane et al. 1997; Damasio 1998; Rossi et al. 2007). The collective Lewy pathology within the central autonomic network can contribute to and exacerbate autonomic changes (Benarroch 1993; Benarroch and Chang 1993; Braak and Del Tredici 2005, 2008).

According to the proposed hypothesis mentioned previously, a pathogen could use short pathways from the central subnucleus of the amygdala to reach the basolateral subnuclei. These nuclei receive powerful input from the temporal mesocortex and, thus, a neurotropic pathogen might well be able to reach this cortical region. At present, the major objection to the idea that axonal interconnectivities might play a major role in the stereotypic progression of the pathological process is the fact that the hypothalamic paraventricular nucleus remains virtually free of Lewy pathology lesions during the entire disease course. This nucleus provides projections to most autonomic relay centers and sends axons directly to the dorsal motor nucleus of the vagal nerve and intermediolateral nucleus of the spinal cord. The exceptional position on the part of the hypothalamic paraventricular nucleus in sPD raises questions for which there are no immediately obvious answers.

8
Stages 5 and 6

In the end stages, the Lewy pathology in the autonomic, limbic, and somatomotor systems is compounded by impairment and functional deficits attributable to LNs/LBs in neocortical areas (Fig. 23a,b). The uninterrupted diminishment of limbic input to the prefrontal cortex can lead to loss of initiative and dementia (Braak et al. 2005; Galvin et al. 2006). The severity of the pathological changes in the sites affected at previous neuropathological stages increases (Figs. 7e,f; 9c,d; 11c; 13e; 17c-f; 18b; 19; 20d,e). The vulnerable portions of the substantia nigra appear under the microscope nearly denuded of melanoneurons and are pale upon macroscopic inspection (Fig. 16g). Severe lesions are seen in the nonthalamic nuclei with diffuse projections as well as in central subnucleus of the amygdala and interstitial nucleus of the terminal stria. Olfactory-related portions of the entorhinal region, periamygdalar cortex, cortical and accessory cortical subnuclei of the amygdala show heavy involvement. With the exception of the medial subnucleus of the amygdala, all components of the olfactory system eventually become affected. Together with the temporal mesocortex, the insular, subgenual, and anterior cingulate regions are among the cortical areas with the highest densities of LB-bearing projection neurons (Fig. 23a-e). By contrast, centers of the striatal circuit containing projection cells with myelin-rich axons, (pallidum, subthalamic nucleus, thalamic relay nuclei) remain intact or nearly intact, even in stage 6 cases. The same holds generally true for most components of the cerebellar circuit (pontine gray, precerebellar brainstem nuclei, inferior olivary nucleus, vestibular nuclei, cerebellar cortex, cerebellar nuclei, parvo- and magnocellular portions of the red nucleus). In exceptional instances, isolated lesions can be seen to develop in portions of these structures (Fig. 28b,c).

8.1 Involvement of Neocortex and Basal Ganglia

8.1.1 Neocortex

The chief characteristic of neuropathological stage 5 is a progression of the lesions from the mesocortex into the adjoining mature neocortex, beginning with the expansive high order sensory association fields of the temporal lobe and prefrontal areas (Figs. 21b, 22). With the transition from the mesocortex to the mature neocortex, the presence of LNs in layers II–III gradually diminishes, and the LN-plexus is no longer observable in layers V–VI. Cortical LBs occur in medium-sized to small pyramidal cells predominantly located in layers V–VI and also, to a lesser extent, in layer IIIc (Wakabayashi et al. 1995). Although the extent of neocortical involvement varies in some individuals, the primary sensory and motor fields, as well as first order sensory association fields and premotor areas are still nearly intact in stage 5 (Fig. 23a), whereas in stage 6 Lewy pathology appears in these areas as well (Fig. 23b). By way of illustration, Fig. 24e–g depicts the gradual diminishment of cortical involvement moving from occipital peristriate areas (Brodmann field 19), through the parastriate field (Brodmann field 18) into the striate area (Brodmann field 17). A similar reduction in lesional density occurs with the transition from the second to the first temporal gyrus and again upon reaching the primary auditory

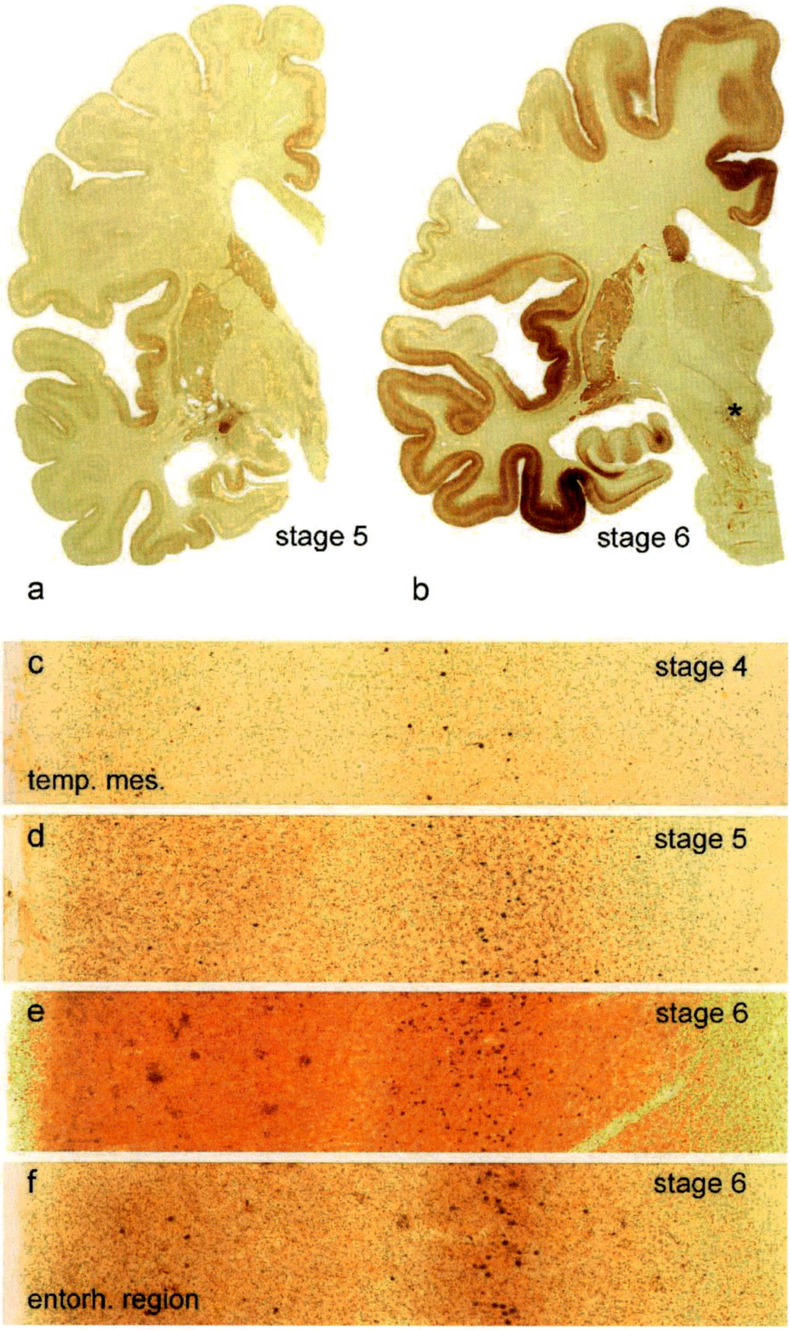
stage 5
stage 6
*
a
b
c
stage 4
temp. mes.
d
stage 5
e
stage 6
f
stage 6
entorh. region

field in the transverse gyrus of Heschl (Figs. 23b, 24d). The sparse distribution pattern of the Lewy pathology in well-myelinated primary fields is similar across most cases at stage 6. There is no evidence in any portion of the cerebral cortex that projection neurons apart from the above-mentioned pyramidal cells are capable of developing LBs; cortical projection neurons originating in other layers and with other destinations do not become involved in sPD.

8.1.2 Basal Ganglia

The basal ganglia lie close to the cerebral cortex and, as such, are optimally positioned to act in conjunction with the cortex. Some striatal circuit centers (e.g., the dorsal striatum, external and internal pallidum, subthalamic nucleus, ventrolateral thalamus) commence myelination prenatally, with the result that the human pallidum and subthalamic nucleus are exceptionally well myelinated. With the exception of the substantia nigra and pedunculopontine tegmental nucleus—whose axons are thin and poorly myelinated—the projection neurons of the pallidum, subthalamic nucleus, and thalamic relay nuclei are resistant to the disease process. Isolated and relatively large LNs begin to appear in the striatum in stage 5, but the identities of the affected nerve cell types are currently unknown. Dot-like immunoreactive inclusions resembling truncated LNs are present since stage 4 in the striatum, claustrum, and in large portions of the thalamic relay nuclei (dotted nuclei

Fig. 23 Lewy pathology in 100 μm polyethylene glycol-embedded hemisphere sections at stages 4, 5, and 6 in immunoreactions against α-synuclein. **a** In stage 5, the severity of the lesions that began in stage 4 worsens, and for the first time the pathology progresses into portions of the neocortex adjoining the mesocortex. **b** The very severe changes seen in cases with stage 6 brain pathology reflect the increasing destruction of the temporal, insular, and cingulate cortex seen in the previous two stages. Note that the differential gradient of cortical involvement (moving clockwise from the anteromedial temporal mesocortex toward the neocortex) persists. Here, the pathological process also has encroached upon first order sensory association fields, reaching even the primary auditory field in the gyrus of Heschl. Observe, too, the gradually increasing pathology in the striatum in neuropathological stages 4, 5, and 6 (compare Fig. 15b with Fig. 23a and Fig. 23b,c–e). Characteristic involvement of the anteromedial temporal mesocortex in stages 4–6. **c** A network of Lewy neurites develops in layers II–III together with Lewy body-bearing pyramidal cells in layers V–VI (stage 4). **d** The lesions increase in severity, and α-synuclein immunoreactive astrocytes appear at stage 5, mainly in cortical layers II–III and V–VI. **e** Very high lesion densities are attained in stage 6. In addition, this section contains Lewy plaques (*large dark rounded areas*; see also Fig. 3l–n). **f** Typical involvement of the allocortical entorhinal cortex (*entorh. region*) at stage 6 with predominant involvement of layer pri-α (the outer layer of the internal principal lamina). *Scale bar* in **c** is also valid for **d–f**

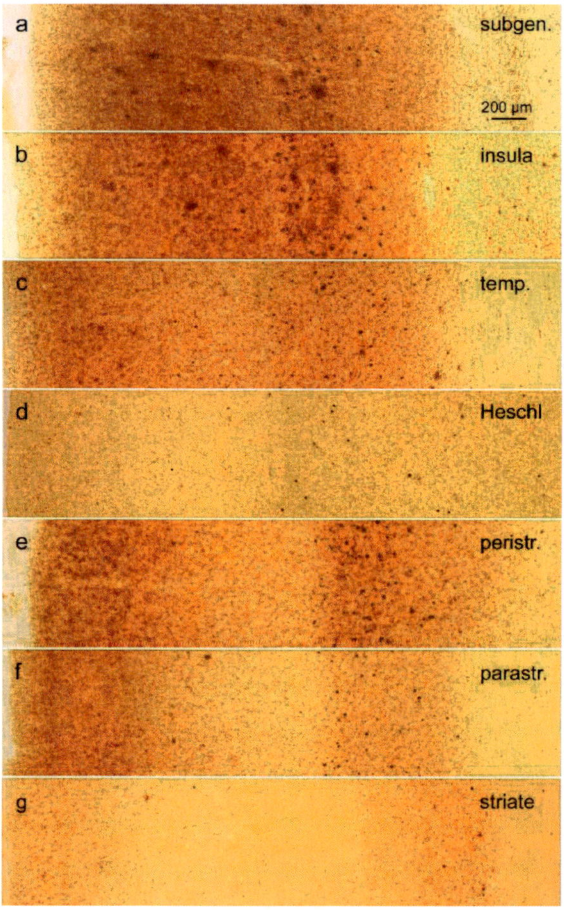

Fig. 24 Lewy pathology in 100 µm polyethylene glycol-embedded hemisphere sections at stage 6 in immunoreactions against α-synuclein. **a, b** Heavy involvement of subgenual (*subgen.*) and insular (*insula*) fields, i.e., cortical visceromotor and viscerosensory limbic integration areas. These areas frequently show the presence of Lewy plaques (*dark rounded areas*). **c, d** Considerable diminution of the Lewy pathology when proceeding from temporal high order association cortex (*temp.*) into the primary auditory field covering the gyrus of Heschl (*Heschl*). **e–g** A similar diminution can be observed, in the occipital lobe, when proceeding from the peristriate high order association region (*peristr.*) through the parastriate first order association area (*parastr.*) into the primary visual field, the striate area (*striate*). *Scale bar* in **a** also applies to **b–g** and is also valid for Fig. 23c–f

in Fig. 29). They become much more noticeable because their numbers and α-synuclein-immunoreactivity increase markedly during stages 5–6 (compare Fig. 23a with 23b).

8.2 Basic Organization of the Basal Ganglia and Striatal Circuit

The mammalian prosencephalic motor system has evolved out of the mesencephalic motor system of nonmammalian vertebrates and reaches its culmination in humans, where motor activity is primarily influenced by the cerebral cortex. Although somatomotor movements are the result of smoothly interacting functional components that extend from the spinal cord to the telencephalon, truly effective motor activity in the human being is dependent to a large extent on the structural integrity and performance of the cortex. The prefrontal cortex and extensive regions of the neocortex initiate movements by activating the dorsal striatum via glutamatergic projections. From there, impulses transferred through the pallidum arrive at the anteroventral thalamic nuclei (VA, VLa), which relay data back to the cortex, thereby establishing the cortico-basal ganglia-thalamo-cortical circuit, which, for the sake of brevity, is referred to here as the striatal circuit (Figs. 27c, 28a; Alexander et al. 1990; Parent and Hazrati 1995a, b; Wichmann and DeLong 2003; Tepper et al. 2007).

8.2.1 Striatum

The striatum comprises the accumbens nucleus/ventral striatum and caudate nucleus/putamen (dorsal striatum). Bundles of myelinated fibers passing through each of these nuclei lend them a striped or "striated" aspect. The accumbens nucleus occupies the central portion of the striatum, forming its anteromedial extremity and generating a superior (caudate nucleus), lateral (putamen), and inferior extension (ventral striatum) (Zahm 2000). The ventral striatum fills much of the space below the anterior commissure and extends downwards to the basal surface of the brain. Together with large portions of the accumbens nucleus, the ventral striatum is closely interconnected with cortical areas and subcortical nuclei of the limbic system (Fig. 22). The head of the caudate nucleus emerges from the accumbens nucleus, narrows posteriorly into the caudate body, and gradually tapers off toward the tail. The caudate nucleus is incompletely separated from the putamen owing to the presence of the intervening fiber bundles of the internal capsule (Holt et al. 1996). Striatal components are uniformly composed of a small number of neuronal types (Parent and Hazrati 1995a, b). More than 90% of the total population are inhibitory GABAergic projection cells represented by the class of medium-sized spiny neurons. The centrally located nucleus of these neurons is pale and sparsely basophilic. The soma contains lipofuscin pigment granules that tend to extend into one or more of the proximal dendrites. Densely arborizing dendrites radiate in all directions: whereas their proximal stems are free of spines, the distal dendritic portions are thickly covered with spines. The axons of these neurons give off extensive local collaterals that terminate mostly on proximal dendrites of nearby projection neurons, thus linking local clusters of such cells (Parent et al. 2000). The axons' terminal branches are sparsely myelinated and, after leaving the striatum, terminate within the pallidum in the form of band-like arborizations oriented parallel to the medullary lamina. Other portions of the axons project to the substantia nigra.

There are two subclasses of striatal medium-sized spiny neurons. The first subclass is characterized by an opiate peptide (mostly an enkephalin-like peptide) and expresses the D2 subtype of dopamine receptors. The second subclass contains substance P, dynorphin, and expresses the D1 dopamine receptor subtype (Figs. 27c, 28a; Alheid et al. 1990). Both classes of nerve cells receive predominantly excitatory glutamatergic afferents from association areas of the cerebral cortex that terminate with asymmetrical contacts on the tips of their dendritic spines (Smith and Bolam 1990). Dopaminergic projection fibers from the substantia nigra form symmetric contacts chiefly on the stalks of the spines, whereas the intrinsic cholinergic and glutamatergic terminals from the thalamus are usually found on the dendritic shafts and proximal stems of the dendrites. Both dopaminergic and cholinergic axon terminals regulate the excitatory striatal input from cortical areas and thalamic nuclei (Smith and Bolam 1990).

Unfortunately, much less is known about the remaining nerve cell types within the striatum. Most, however, appear to function as local circuit neurons (Kawaguchi et al. 1995; Tepper and Bolam 2004). One relatively well-defined class of local circuit neurons is that of the cholinergic aspiny cells. These neurons have multiple, thin winding dendrites and a short axon that arborizes extensively into terminal branches (Yelnik et al. 1991). The rounded and comparatively voluminous cell bodies exhibit an excentrically placed small nucleus opposite a tightly packed accumulation of lipofuscin pigment granules and peripherally located Nissl material. The nuclear envelope facing the pale center of the cell body has numerous deep folds. These aspiny striatal cells receive cortical glutamatergic afferents and generate most of the cholinergic fibers within the striatum (Fig. 28a). In addition, they have an excitatory effect on striatal GABA/enkephalin projection neurons (Calabresi et al. 2000). Other subclasses of local circuit neurons that populate the striatum are the parvalbumin-positive GABAergic cells, the somatostatin- and neuropeptide Y (NPY)-immunoreactive neurons, and the calretinin-positive cells (Parent and Hazrati 1995a; Tepper and Bolam 2004).

The dorsal striatum receives topographically organized glutamatergic projections from all areas of the neocortex, and these afferents originate from pyramidal projection neurons in both the supra- and infragranular layers (van Hoesen et al. 1981; Yeterian and Pandya 1991, 1994; Parent and Hazrati 1995a). Approximately 80% of all synapses in the striatum are derived from these corticostriatal projections (Fig. 28a; Pasik et al. 1976). Highorder association fields of the neocortex project to the caudate nucleus, whereas the sensorimotor cortex of the central region targets mostly the putamen with a somatotopic representation of the body in the form of obliquely oriented stripes (Selemon and Goldman-Rakic 1985). An additional topographically organized glutamatergic projection toward the dorsal striatum originates in the thalamic intralaminar nuclei. Another source of major striatal input is the dopaminergic projection from the substantia nigra (pars compacta). In the adult human, dopamine terminals are densely and homogeneously distributed through the striatum. Less dense projections to the striatum originate from GABAergic histamine-containing nerve cells of the hypothalamic tuberomamillary nucleus, from serotonergic neurons of the raphe nuclei, and from noradrenergic neurons of the coeruleus-subcoeruleus complex (Halliday 2004). Projection fibers from the dorsal striatum, on the other hand, approximate a roughly somatotopic topography and terminate within the pallidum and substantia nigra.

8.2.2 Pallidum

The diencephalic pallidum makes up the most anterior part of the hypothalamus and derives its name from the pale appearance lent to this nucleus by its myelinated fibers in fresh material. The external segment of the dorsal pallidum is separated from the convexity of the putamen by the external medullary lamina, and a corresponding internal lamina sets off the internal segment from the external one. The pallidum's posterior extremity is contiguous with the reticulate portion of the substantia nigra. The boundaries of the ventral pallidum are less clearly defined. Anteromedially it is situated close to the interstitial nucleus of the terminal stria and the thalamic reticular nucleus. Inferiorly it abuts on the nucleus of the diagonal band and hypothalamic tuberomamillary nucleus. The ventral pallidum merges into the lateral hypothalamus and extends beneath the anterior commissure. Both compartments of the pallidum are populated by elongated nerve cells that are spaced widely apart. The somata generate stout and extensive, sparsely spined, and only occasionally arborizing dendrites, which are oriented parallel to the medullary laminae and perpendicular to the incoming striatal axons that surround the elongated nerve cells in the form of a bushy plexus. The large GABAergic nerve cells of the pallidum have few lipofuscin pigment deposits and strongly basophilic Nissl bodies. The glial cells and neurons within the pallidum and within the related nigral reticulate zone contain endogenous iron (Alheid et al. 1990). The external and internal pallidal segments as well as the reticulate zone of the substantia nigra receive topographically arranged afferents from the dorsal striatum. Whereas GABA/enkephalin projections terminate preferentially in the external segment, the internal segment receives GABA/substance P fibers (Beach and McGeer 1984; Reiner and Anderson 1990). The external pallidum provides a dense projection to the subthalamic nucleus and sparser projections to the striatum, internal pallidum, nigral reticulate zone, and thalamic reticular nucleus (Fig. 28a; Hazrati and Parent 1991; Parent and Hazrati 1995b; Chesselet and Delfs 1996; Smith et al. 1998).

The main feature of the external pallidum is that it establishes, via the subthalamic nucleus, an "indirect" pathway to the main pallidal output structure—the internal segment (Figs. 27c, 28a). Under resting conditions, the activity of striatal spiny projection neurons is minimal. Activation of the inhibitory GABA/enkephalin nerve cells results in suppression of projection cells within the external pallidum, thereby disinhibiting or activating the subthalamic nucleus, which in turn activates the internal pallidum (Fig. 27d). The inhibitory influence of the internal pallidum eventually suppresses thalamic activity (mainly VA and VLa) and provides a negative feedback to the cortex (Fig. 27d). Increased activity of the skeletal musculature, on the other hand, requires that striatal circuitry centers involved in the "direct" pathway predominate; activation of inhibitory striatal substance P neurons suppresses part of the activity in the internal pallidum and results in the activation of related thalamic nuclei, thereby opening the way for positive feedback to the cortex (Figs. 27e). It should be noted that the two populations of striatal projection neurons with their different pallidal recipients have opposing effects on cortical motor activity (Fig. 27d,e; Alexander et al. 1990; Graybiel et al. 1990; Gerfen 1992; Albin et al. 1995).

The dopaminergic projection from melanized neurons in the substantia nigra appears to exert an opposing influence on the aforementioned indirect and direct striatal pathways by inhibiting the indirect pathway via D2 receptors of GABA/enkephalin neurons and activating the direct pathway via D1 receptors of GABA/substance P neurons (Starr 1995; Gerfen 2000). Activation of striatal substance P neurons results in the temporary dominance of the centers belonging to the direct pathway, as befits the prompt execution of the incoming motor impulses from the neocortex. The temporary inhibition of striatal enkephalin neurons compounds this activating effect. By promoting the direct pathway and curbing the indirect one, the neurotransmitter dopamine facilitates the passage of data through the striatal circuit and paves the way for the rapid and smoothly functioning stimulation of cortical motor areas (Fig. 28). A clear-cut separation of the two pathways in the limbic basal ganglia circuitry does not exist (Fig. 22). The anatomical organization of the striatal circuit depicted in Fig. 27c is intended to serve as a basis for understanding the motor dysfunctions (hypokinesia, bradykinesia) that develop in the course of sPD (Alexander et al. 1990; DeLong 1990; Albin et al. 1995; Wichmann and DeLong 1993, 1997, 1998, 2003; Obeso et al. 1997; Thiel et al. 2003).

8.2.3 Subthalamic Nucleus

The lentil-shaped subthalamic nucleus (body of Luys) is situated above the internal capsule and cerebral peduncle. Posteriorly it superimposes on the pars compacta of the substantia nigra. The nucleus consists of a single type of medium-sized nerve cell that has relatively long dendrites with irregularly spaced spines (Yelnik and Percheron 1979). The axons give off local collaterals and the main axonal trunk leaves the nucleus to terminate in the internal pallidum and reticulate portion of the substantia nigra. One of the main attributes of the subthalamic nucleus is that it receives remarkably heavy input directly from the cerebral cortex—in particular, from the primary motor field and, to a lesser extent, from premotor areas (Figs. 28a, 29c; Hamani et al. 2004). It also receives afferents from both compartments of the pallidum as well as from the pedunculopontine tegmental nucleus, and it sends a powerful excitatory glutamatergic projection to the internal pallidum (Figs. 28a, 29). This combination of interconnectivities makes the subthalamic nucleus one of the driving forces within the basal ganglia system. The nucleus also provides a feedback projection to the external pallidum (Figs. 28a, 29). Finally, sparse subthalamic projections reach the striatum, the pedunculopontine nucleus, and spinal cord (Smith et al. 1998).

8.2.4
Claustrum

The claustrum is a subcortical nucleus located beneath the insular cortex and is separated from it by the extreme capsule. It maintains reciprocally organized connectivities to virtually all portions of the cerebral cortex. The morphological features of claustral projection neurons resemble those of the pyramidal cells in layers V and VI of the neocortex. Claustral projection neurons give rise to long axons that terminate chiefly in layer IV and, to a lesser extent, in layer VI. Among the cortical neurons that generate projections to the claustrum are small pyramidal cells located in layer VI. The anterior cingulate areas, entorhinal region, and subiculum project to ventral portions of the claustrum, and bilateral projections connect the ventral claustrum with the intralaminar nuclei of the thalamus (Sherk 1986). Thus, the claustrum—similar to the intralaminar nuclei—assumes the functions of a diffusely projecting relay between autonomic centers and the entire cerebral cortex (Fig. 29). In stages 5 and 6, Lewy pathology appears mainly within the claustrum's ventral portion: spherical LBs of the cortical type occur solely in the somata of claustral projection neurons. Short and thin LNs usually devoid of arborizations accompany the involved projection cells.

8.3
Dot-Like Inclusions and Abnormally Altered Astrocytes in the Forebrain

Small dot-like α-synuclein-immunoreactive inclusions, probably aggregations of the protein in terminal axons, first appear at stage 4 in layers II–III and V–VI of the anteromedial temporal mesocortex, in the striatum (excluding the pallidum and subthalamic nucleus), thalamic relay nuclei, claustrum, and amygdala (Braak et al. 2007b). These dot-like particles increase noticeably in density during stages 5 and 6 (dotted nuclei in Fig. 29). In the striatum, most of the inclusions are seen in neuronal processes that contain neurofilament subunits but not in those that are immunopositive for the dendritic marker microtubule-associated protein (MAP)-2 (Duda et al. 2002). Occasionally, dot-like aggregates can be observed in tyrosine hydroxylase-immunoreactive nerve cell processes that may represent axons of dopaminergic nigral projection neurons (Duda et al. 2002). Closer inspection of the punctate forebrain lesions in sPD, however, reveals that not only nerve cells and their axons but also nonneuronal cells are involved (Fig. 25a; Braak et al. 2007b). Previous studies have reported the occurrence of α-synuclein-inclusions in glial

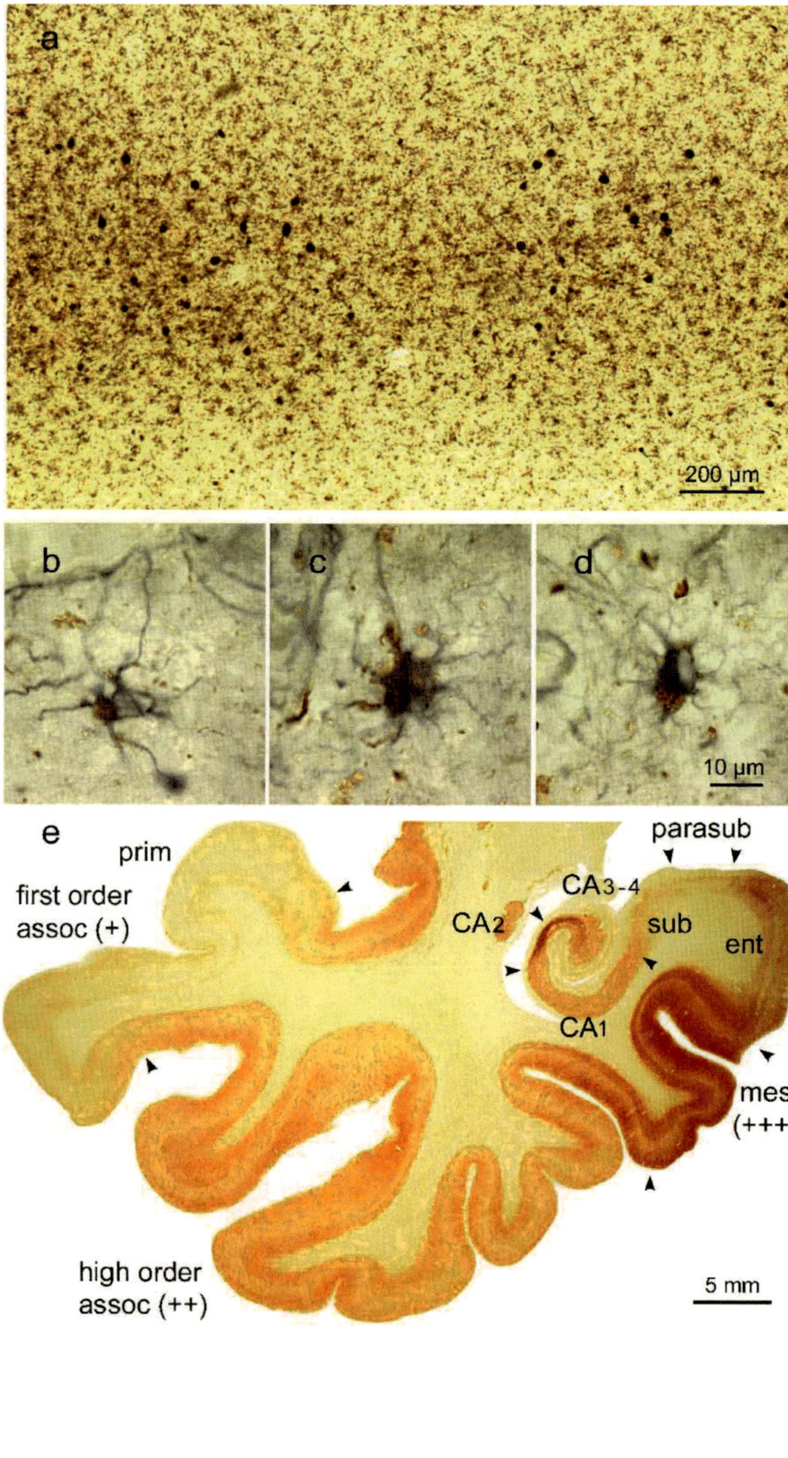
a
200 µm
b
c
d
10 µm
e
prim
first order
assoc (+)
parasub
CA3-4
CA2
sub
ent
CA1
mes
(+++)
high order
assoc (++)
5 mm

cells (Takeda et al. 1998, 2000; Arai et al. 1999; Piao et al. 2000; Hishikawa et al. 2001, 2005; Togo et al. 2001; Mori et al. 2002; Wakabayashi et al. 2000b; Terada et al. 2003), but most refer to abnormal aggregations of this protein in oligodendroglial cells (Arima et al. 1998b; Gai et al. 1998; Tu et al. 1998). Astrocytic α-synuclein inclusions in sPD, on the other hand, are described as being argyrophilic (Arai et al. 1999; Hishikawa et al. 2001). The nonneuronal cells illustrated in Fig. 25b–d are nonargyrophilic (i.e., Gallyas-, and Campbell–Switzer-negative; Braak et al. 2007b). In sPD, astrocytic inclusions appear in the temporal mesocortex, neocortex, striatum, and thalamic relay nuclei in that order. Morphologically, the size, aspect, and characteristics of these cells, including their faint cell nuclei and sparse nonbasophilic cytoplasm, make it possible to identify them as astrocytes. The fact that they exhibit GFAP-immunoreactivity (Fig. 25b–d) also corroborates this finding.

Astrocytic α-synuclein aggregations appear to contain a homogeneous particulate material that accumulates near the cell nucleus and extends, with decreasing packing density, into the proximal cellular processes. Pretreatment with formic acid and use of a primary antibody that recognizes residues 91–99 within the NAC region of the α-synuclein molecule consistently produce strong immunolabeling of such astrocytic inclusions (Fig. 25b–d; Arima et al. 1998a; Mori et al. 2002; Takeda et al. 1998, 2000). The same aggregations, however, remain unlabeled by antibodies that recognize portions of the α-synuclein molecule outside the NAC region. Notably, the abnormal material in astrocytes is ubiquitin-negative and negative for antibodies against p62, an ubiquitin-binding and signaling protein (Kuusisto et al.

Fig. 25 **a** Deep layers V and VI of the temporal neocortex at stage 5 (100 μm polyethylene glycol-embedded section, immunoreaction against α-synuclein). Following pretreatment with formic acid, the antibody syn-1 (recognizing amino acids 91–99 of human α-synuclein) robustly labels the large intraneuronal Lewy bodies as well as immunoreactive astrocytes in the vicinity of involved pyramidal cells. **b–d** Cortical astrocytes from layers V–VI of the temporal neocortex at stage 5 (6-μm-thick paraffin section). α-Synuclein aggregates are shown in an immunoreaction, as described above, together with an antibody against glial fibrillary acidic protein (GFAP) to identify them as astrocytes. **e** Tissue section through the temporal lobe cut in the frontal plane at the latitude of the lateral geniculate body (100 μm, immunoreaction against α-synuclein). The hemisphere shows allocortical regions (*CA1–4*, sectors of the Ammon's horn; *sub*, subiculum; *parasub*, parasubiculum; *ent*, entorhinal region) together with portions of the anteromedial temporal mesocortex (*mes*), high order sensory association areas (*assoc*), first order sensory association areas, and primary auditory field (*prim*) in the transverse gyrus of Heschl. *Arrowheads* mark the approximate limits of each region. Remarkable are the differences in the density of the immunoreaction, which ranges from strong immunolabeling (+++) in the temporal mesocortex to faint immunopositivity (+) in the primary auditory field of the neocortex (see also Fig. 24c,d and e–g). Neuronal involvement is mainly confined to the deep layers V–VI, whereas the abnormal astrocytic reaction includes deep (V–VI) and superficial (II–III) layers alike. The intensity of the astrocytic immunoreactivity parallels the severity of the neuronal involvement in sporadic Parkinson's disease. By contrast, the white matter remains uninvolved. *Scale bar* in **d** is also valid for **b** and **c**

2003). Immunoreactive astrocytes do not occur in the cortex of controls or in cases with stages 1–3 Lewy pathology. Between stages 4–6, they appear to be confined to specific forebrain sites, including the amygdala, claustrum, striatum, thalamic relay nuclei, intralaminar nuclei of the thalamus, and cortical areas (compare Fig. 15a,b with Fig. 23a,b).

By contrast, not only the subthalamic nucleus but also the midbrain, lower portions of the brainstem, cerebellum, and spinal cord remain free or nearly free of α-synuclein-labeled astrocytes (Braak et al. 2007b). The astrocytic reaction gradually increases in stages 5–6 and keeps pace with the growing degree of cortical neuronal involvement (Fig. 25e). It is most pronounced in the deep layers V–VI (Fig. 25a), grows weaker in layers II–III, and there is little astrocytic immunoreactivity in layers I and IV (Fig. 25e). In affected forebrain regions, the general impression is that practically all astrocytes display immunolabeling and not only those in the vicinity of nerve cells with LBs. Labeled astrocytes continue to maintain their normal distance from each other, and the intensity of the astrocytic immunoreaction gradually tapers off at the limits of uninvolved sites, e.g., in layer IV or in deep portions of the white substance (Fig. 25e).

Closer scrutiny of the striatum and thalamic relay nuclei reveals astrocytic involvement there as well (Fig. 26a,d). Apparently, altered astrocytes only develop in the striatum and thalamus when LBs are present in cortical pyramidal cells, i.e., between stages 4 and 6 (Braak et al. 2007b). Figure 26b,c,e, and f show the astrocytes as granular structures of similar size and regularly spaced. In the striatum and thalamus, immunoreactive dots (probably located in pathologically changed axon terminals) are intermingled with the larger abnormally appearing astrocytes.

Astrocytes are not known to synthesize the protein α-synuclein (Mori et al. 2002). Inasmuch as the physiologically soluble α-synuclein is continuously released into the extracellular space and can be detected in cerebrospinal fluid and peripheral blood plasma (Borghi et al. 2000; El-Agnaf et al. 2003, 2005; Lee et al. 2005; Sung et al. 2005; Mollenhauer et al. 2008), it is conceivable that the protein, in a still soluble state, may be taken up by astrocytes and degraded. After all, even gray matter astrocytes, such as those in the striatum and in layers V–VI of the cerebral cortex (where the neurons have large stores of α-synuclein), as a general rule do not develop α-synuclein inclusions. It is much more likely that the α-synuclein molecule that contributes to the formation of the astrocytic aggregations in sPD originates from nerve cells but is subjected to abnormal

Fig. 26 (continued) α-synuclein throughout the thalamus. The posterolateral ventral nucleus (*VLp*) as well as the laterodorsal (*LD*) and mediodorsal (*MD*) nuclei display relatively strong immunostaining, whereas the immunolabeling in the posteromedial ventral nucleus (*VPm*) is weak, and the centromedial (*Cm*) and parafascicular (*Pf*) nuclei appear nearly devoid of normal α-synuclein. **e, f** Section through the posterolateral ventral nucleus (*VLp*) near the white substance. Closer inspection reveals that the α-synuclein material is mainly concentrated in astrocytes (dark patchy areas) and in small punctate processes in between the astrocytes

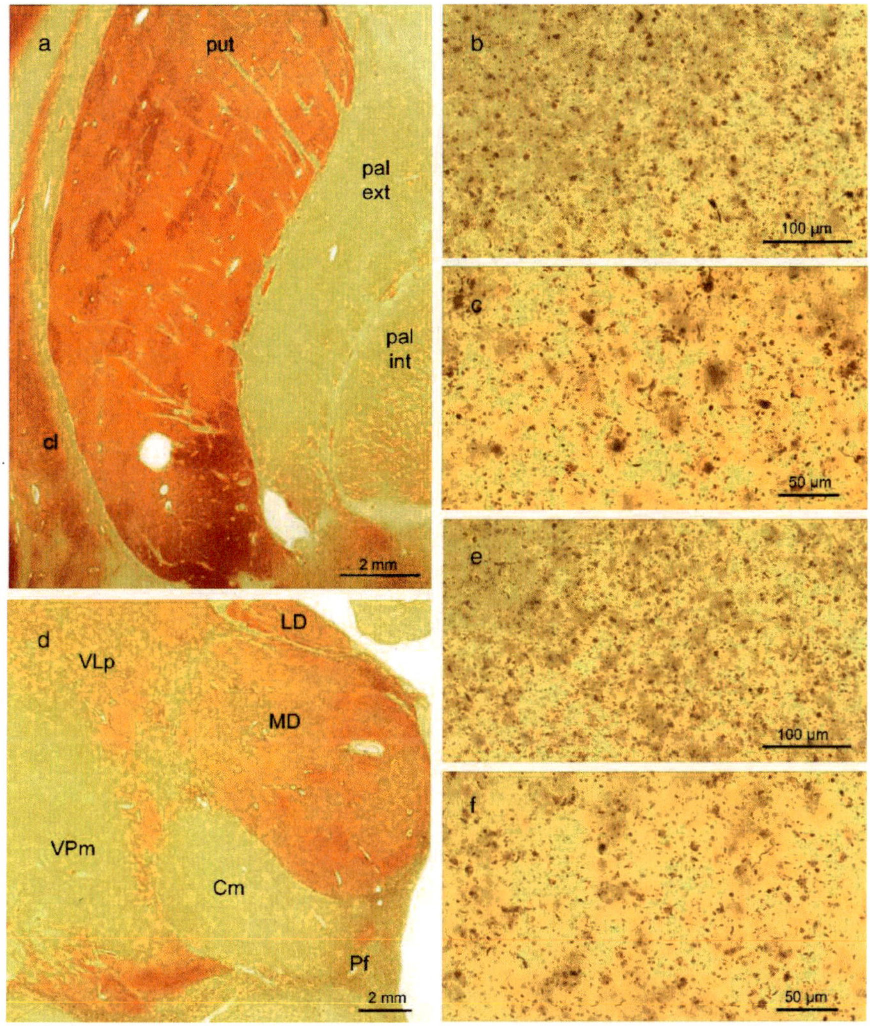

Fig. 26 Lewy pathology in the striatum and thalamus at stage 6 in 100 µm polyethylene glycol-embedded hemisphere sections immunostained for α-synuclein. **a–c** Overview and details of the putamen. Observe the dense immunoreaction in both the putamen (*put*) and claustrum (*cl*) as well as the relative paucity of α-synuclein in external and internal portions of the pallidum (*pall ext* and *pall int*), regardless whether the protein appears in soluble form or in aggregated inclusion bodies. **b, c** On closer inspection, large numbers of α-synuclein-immunoreactive astrocytes as well as punctate structures are visible in the neuropil of the putamen and caudate nucleus (not shown). Astrocytes continue to maintain their normal distance from each other, and the aggregated intra-astrocytic material (large patchy areas in c) can easily be distinguished from the small punctate structures shown at higher magnification in **c**. **d–f** Overview and details of the thalamus. **d** Note the uneven distribution of normal

processing, such as protein misfolding, possibly prior to its release from slightly damaged axon terminals of involved cortical neurons (Lee et al. 2005). Obviously, the amount of α-synuclein "lost" would not be sufficient to prevent the affected neurons from ultimately developing LBs, but it might partially explain why cortical LBs, in comparison to those of the brainstem type, lack a peripheral halo (den Hartog Jager and Bethlem 1960; Forno 1996; Spillantini et al. 1997; Braak et al. 2007b). Inasmuch as the GABAergic projection neurons of the striatum and the glutamatergic nerve cells of thalamic relay nuclei as well as those of the subthalamic nucleus remain free of LBs for the entire duration of the disorder, it is difficult to account for the astrocytic reaction in the thalamus and striatum, were it not for the existence of strong interconnectivities between the cerebral cortex and these nuclei (Fig. 27a). The corticostriatal and corticothalamic pro-

Fig. 27 **a** Diagram of corticostriatal and corticothalamic connectivities (axons in red). It appears that astrocytes respond secondarily to the Parkinson's disease-related involvement of defined forebrain projection neurons. Provided that immunoreactive Lewy body-containing cortical pyramidal neurons in layers V–VI project to affected portions of the striatum and thalamus, it is conceivable that the immunoreactive material in astrocytes there may be caused by a slightly altered form of the protein α-synuclein that "leaks" from terminal axons of involved cortical pyramidal cells (or involved projection cells in the intralaminar thalamic nuclei). The altered proteinaceous material could then be taken up by surrounding astrocytes (arrows) and would account for the presence of the immunolabeled intra-astrocytic inclusion bodies. It is also remarkable that astrocytes located at greater distances from such terminal axons (e.g., astrocytes in layer IV of the neocortex, in the white substance, or in the pallidum) remain intact, i.e., do not develop the abnormal α-synuclein aggregations. The same holds true for the uninvolved or very mildly involved primary motor area. It mainly projects to the subthalamic nucleus, which, accordingly, also remains devoid of α-synuclein-immunoreactive astrocytes. **b** Diagram of the pathological process during neuropathological stages 1–3. The disease process commences in the dorsal motor nucleus of the vagal nerve (stage 1, *black shading*). Premotor and motor neurons of the spinal cord and brainstem, by contrast, remain uninvolved for the duration of the disorder. Lewy pathology then appears within the gain setting nuclei (i.e., lower raphe nuclei, gigantocellular reticular nucleus, coeruleus-subcoeruleus complex) in the lower brainstem (stage 2, *dark gray shading*). These nuclei send descending projections to somatomotor and visceromotor neurons and become involved prior to the substantia nigra, pars compacta. Thereafter, relay centers of the visceromotor system (central subnucleus of the amygdala) and nuclei influencing the somatomotor system (pedunculopontine tegmental nucleus, dopaminergic neurons of the substantia nigra) are drawn into the disease process (stage 3, *light gray shading*). Both the pedunculopontine nucleus and central nucleus of the amygdala send descending projections to the gain setting nuclei and project to all of the nonthalamic nuclei with diffuse cortical and subcortical efferents, which likewise become involved in stage 3. **c–e** Basic model of the direct and indirect pathways through the cortico-basal ganglia-cortical circuit. Terminal axons of midbrain nigral dopaminergic neurons activate substance P/GABAergic (*subst. P*) projection cells in the dorsal striatum while inhibiting enkephalin/GABAergic (*enkeph.*) neurons. Striatal outflow reaches neocortical motor areas via relay nuclei of the ventral thalamus (*VA/VLa*). Abbreviations: *D*1, *D*2, dopamine 1 and dopamine 2 receptors; *VA*, ventral anterior nucleus of the thalamus; *VLa*, anterior ventrolateral nucleus of the thalamus

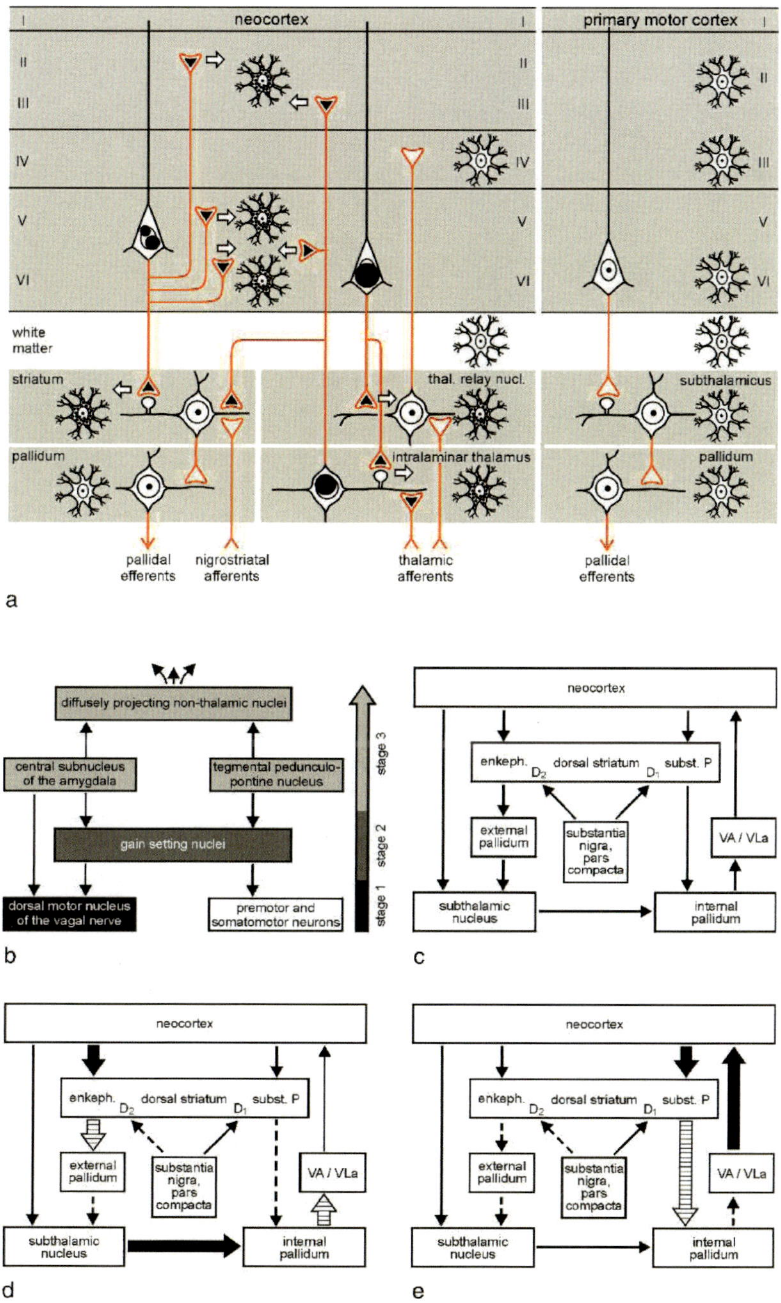
neocortex
primary motor cortex
I
II
III
IV
V
VI
white matter
striatum
pallidum
thal. relay nucl.
intralaminar thalamus
subthalamicus
pallidum
pallidal efferents
nigrostriatal afferents
thalamic afferents
pallidal efferents
a
diffusely projecting non-thalamic nuclei
central subnucleus of the amygdala
tegmental pedunculo-pontine nucleus
gain setting nuclei
dorsal motor nucleus of the vagal nerve
premotor and somatomotor neurons
stage 3
stage 2
stage 1
b
neocortex
enkeph. D2 dorsal striatum D1 subst. P
external pallidum
substantia nigra, pars compacta
VA / VLa
subthalamic nucleus
internal pallidum
c
d
e

jection neurons are small to medium-sized pyramidal cells that primarily reside in layers V–VI (Royce 1982; Jones 1984), and this description corresponds precisely to that of cortical nerve cells that develop Lewy pathology (Fig. 25a). The subthalamic nucleus also receives a dense and heavily myelinated projection from the cerebral cortex, albeit mainly from the primary motor field (Fig. 27a). These circumstances would seem to indicate that the immunoreactive material in astrocytes may originate from material produced in affected cortical nerve cells. Nonstaining of the subthalamic nucleus during the immunoreaction lends credence to this explanation because the parent cells of the corticosubthalamic tract are located in the primary motor field and remain uninvolved or react only occasionally and extremely late in the course of sPD.

8.4 Disruption of the Striatal Circuit

Depletion of dopamine in the striatum probably induces postsynaptic changes in striatal projection neurons and causes an imbalance in the activity within both the direct pathway (insufficient activation) and the indirect pathway (insufficient inhibition). The net effect of the inequilibrium is hyperactivity or disinhibition of the subthalamic nucleus. The resultant predominance of the internal pallidum leads, in turn, to inhibition of thalamocortical activity and, clinically, to hypokinesia or akinesia (Wichmann and DeLong 1998; DeLong and Wichmann 2007). Pharmacological alleviation of the striatal dopamine deficiency or a surgically induced reduction of the hyperactivity in the subthalamic nucleus (deep brain stimulation) can restore, to an extent, equilibrium between the indirect and direct pathways (Figs. 27c, 28a, Benabid 2003; Kopell et al. 2006; Temel and Visser-Vandewalle 2006; Wichmann and DeLong 2006; Stefani et al. 2007; Volkmann 2007). The classical scheme of the striatal circuit (Fig. 27c), however, has several limitations (Albin et al. 1995). It suggests that striatal output reaches the internal pallidum via two nearly separate or independent pathways—an oversimplification because both striatal projections, alone owing to the existence of extensive local collaterals, are known to

Fig. 28 **a** Amended diagram of the corticobasal ganglia-cortical circuit. This version encompasses motor areas from the spinal cord to the neocortex and incorporates not only the consequences of dopamine depletion in the dorsal striatum but also the influence of nondopaminergic centers that become consecutively and severely impaired in sporadic Parkinson's disease. Cortical Lewy pathology most probably impairs the corticostriatal projection (*large arrowheads*), whereas the corticosubthalamic connection remains intact. For a fuller explanation, see Sect. 8.4. **b, c** Overview (**b**) and detail (**c**) of the Lewy pathology that occasionally develops in the precerebellar nuclei (here the principal nucleus of the inferior olive) in late stages 5 or 6. **Abbreviations:** *ACh*, acetylcholine; *D1*, *D2*, dopamine 1 and dopamine 2 receptors; *VA*, ventral anterior nucleus of the thalamus; *VLa*, anterior ventrolateral nucleus of the thalamus

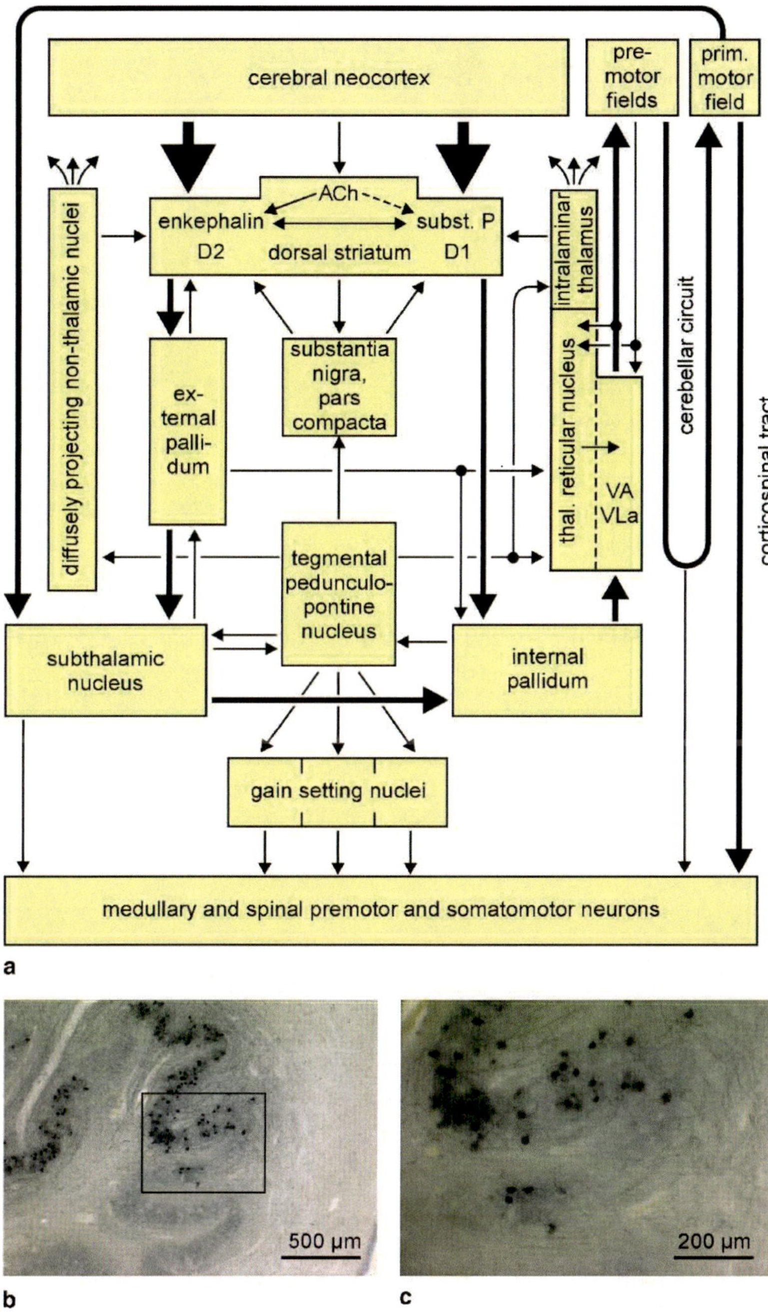
cerebral neocortex
pre-motor fields
prim. motor field
diffusely projecting non-thalamic nuclei
ACh
enkephalin
subst. P
D2
dorsal striatum
D1
intralaminar thalamus
ex-ternal palli-dum
substantia nigra, pars compacta
thal. reticular nucleus
VA
VLa
cerebellar circuit
corticospinal tract
tegmental pedunculo-pontine nucleus
subthalamic nucleus
internal pallidum
gain setting nuclei
medullary and spinal premotor and somatomotor neurons
a
500 µm
b
200 µm
c

be reciprocally more closely connected than previously thought (Yung et al. 1996). Amended versions of the basic model have proposed more refined organization of the striatal circuit (Chesselet and Delfs 1996; Parent and Cicchetti 1998; Blandini et al. 2000). Closer consideration of the efferent projections of the external pallidum, for instance, reveals that it projects to the reticular nucleus of the thalamus and sends data directly to the internal pallidum without interposition of the subthalamic nucleus (Fig. 28a). The basic model also offers no satisfactory explanation for why unilateral lesions of centers belonging to the direct pathway (internal pallidum, VA/VLa) do not result in an exacerbation of motor dysfunctions (Marsden and Obeso 1994). By the same token, lesioning of the subthalamic nucleus should result in disinhibition of the related thalamic nuclei, thereby producing hyperkinesia, dyskinesia, or both, and yet the postoperative results of stereotactic lesioning show the opposite to be the case (Blandini et al. 2000; Betchen and Kaplitt 2003).

Possibly the most important and heretofore overlooked abnormal change in the striatal circuit during sPD is the considerable loss of spines that occurs along the dendrites of striatal projection neurons (McNeill et al. 1988; Lach et al. 1992; Stephens et al. 2005; Zaja-Milatovic et al. 2005; Kramer and Schulz-Schaeffer 2007). Spine loss, however, is preceded by a depletion of striatal dopamine, which at first leads to hyperactivity on the part of striatal projection neurons. Reduction of glutamate receptors or a loss of spines would both be appropriate mechanisms for curbing corticostriatal glutamatergic drive. Indeed, many authors assume that, to a certain extent, striatal projection neurons eliminate their dendritic spines as a plastic or protective response to dampen excessive cortical input and to ensure their own survival (Day et al. 2006; Deutch 2006; Gerfen 2006; Neely et al. 2007). At any rate, levodopa therapy does not appear to reverse the process.

Although the loss of spines has been ascribed chiefly to the sPD-associated reduction of dopaminergic input, the presence of the axonal dot-like inclusions and α-synuclein immunoreactive astrocytes in the striatum permits another, and more plausible, interpretation. It is likely that cortical Lewy pathology develops from stage 5 onwards within pyramidal cells that generate the corticostriatal and corticothalamic projections. Abnormal dot-like inclusions in their terminal axons would result in a successive loss of glutamatergic synapses that ultimately renders the spines useless. Most of the synapses in the striatum belong to the corticostriatal projection and, for this reason, impairment or destruction of cortical glutamatergic synapses, i.e., the dominant input to the striatum, can be anticipated to severely interrupt the data stream directly at the "doorstep," as it were, of the striatal circuit (Fig. 28a). Damage at this juncture is accompanied by lesions within corticothalamic projections at the other end of the striatal circuit (Fig. 28a). The observation that patients eventually become refractory to the therapeutic benefits of levodopa can partially be accounted for by striatal dopamine depletion, but the reduced response is also attributable to the progressive impairment of the corticostriatal pathway, which gains in importance during the final phase of the disease.

Finally, the hyperactivity that develops in the subthalamic nucleus is also in need of an explanation. Subthalamic disinhibition most probably does not result exclusively from the disinhibitory influence of the external pallidum, as intimated by the basic model (Fig. 27c; Levy et al. 1997; Blandini et al. 2000). The hyperactivity might also result from the excitatory influence of corticosubthalamic connectivities, which mainly are generated from the primary motor area (Smith et al. 1998). In sPD, the primary motor field remains intact, and, accordingly, the corticosubthalamic tract is fully functional even in the end stages of the disorder. On the other hand, the corticostriatal projection forfeits its excitatory input, so that the subthalamic nucleus is increasingly subjected to the activating influence of the primary motor cortex. This interpretation not only accounts for the hyperactivity of the subthalamic nucleus but also assigns the nucleus a more self-contained and prominent role in the organization of the striatal circuit than previously.

In Fig. 28, an attempt has been made to go beyond the treatment of striatal circuit components in isolation and instead to place the striatal circuit within the context of the larger pathological process by incorporating the most important somatomotor centers of the CNS. The new model is simplified for the sake of clarity and extends from the cerebral cortex to the premotor and motor neurons of the brainstem and spinal cord. This amended version of the basic model assigns a pivotal position to the pedunculopontine tegmental nucleus, which does not appear in the classical scheme of the striatal circuit (compare Fig. 27c with Fig. 28a). The pedunculopontine tegmental nucleus receives its major afferents from the internal pallidum and subthalamic nucleus, and sends efferents both to nigral dopaminergic neurons and to the nuclei of the gain setting system in the lower brainstem. In this manner, the pedunculopontine nucleus probably establishes a descending outflow pathway from the otherwise closed cortico-basal ganglia-cortical circuit and directs portions of the data stream via lower brainstem nuclei to premotor and motor neurons (Lee et al. 2000; Nakano 2000; Matsumura and Kojima 2001; Mena-Segovia et al. 2004). Given its superordinate influence on somatomotor activity, the pedunculopontine nucleus is analogous to the central subnucleus of the amygdala that primarily influences, likewise via the gain setting complex, visceromotor activity, and both nuclei send efferents to the nonthalamic nuclei with diffuse projections (Fig. 27b). The potential consequence of the nondopaminergic lesions in both nuclei have received too little attention to date (Sethi 2008). In this context, it should be emphasized that the nuclei of the gain setting system are drawn into the disease process in stage 2, so that impairment of this complex alone could possibly suffice to induce mild but specific dysfunctions within the somatomotor system (Loza et al. 1997; Valls-Solé 2000; Valls-Solé and Valldeoriola 2002). These complaints would become exacerbated, and new symptoms appear, following the involvement of the pedunculopontine tegmental nucleus and substantia nigra in stage 3. Involvement of the thalamic intralaminar nuclei follows in stage 4, and increasing reduction of the corticostriatal and corticothalamic connectivities occurs in stages 5–6. The anticipated result would be severe deterioration of the somatomotor system.

8.5
Potential Functional Consequences of Lesions

Reference to the somatomotor dysfunctions that develop in the course of sPD has already been made in Sect. 8.2. Nevertheless, above all in the so-called "motor phase" of sPD, it is not always the motor symptoms that are in the foreground. Patients, their families, and their caregivers often are more troubled by the development of debilitating and therapy-resistant nonmotor symptoms, such as the gradually increasing impairment of cognitive abilities. All of the high-order limbic circuit centers (entorhinal region, hippocampal formation, amygdala) and many of the cortical fields and subcortical nuclei connected with them sustain heavy neuronal damage in stages 5 and 6 (Figs. 22, 29). It should again be emphasized here that the preeminent organizational level of the human limbic system is essentially neocortex-oriented and not only receives sensory data via the anteromedial temporal mesocortex from neocortical parietal, occipital, and temporal areas but also sends powerful projections to the prefrontal neocortex (Fig. 22). As pointed out above, the maximal cortical lesional density in sPD is seen in the anteromedial temporal mesocortex. The second point of entry for neocortical data into the limbic circuit, the lateral nucleus of the amygdala, also exhibits pronounced Lewy pathology. Finally, on the efferent side, neuronal loss is found in the basal and accessory basal nuclei of the amygdala that generate projections to the ventral striatum, ventral pallidum, mediodorsal thalamus, and prefrontal cortex. In short, the Lewy pathology in the anteromedial temporal mesocortex and amygdala undermines the data transfer from the sensory neocortex via the entorhinal region and amygdala to the prefrontal cortex (Figs. 22).

Other neurodegenerative disorders that inevitably lead to dementia, such as Alzheimer's disease, stand out owing to the massive intraneuronal pathology and neuronal loss in the cerebral cortex—features that make it relatively easy to explain the functional damage and clinical symptoms. In sPD, on the other hand, the cerebral cortex remains intact until neuropathological stage 4. Larger portions of the neocortex initially become involved in the following stage, but it is in stage 6 that the inclusion body pathology attains truly striking proportions. Moreover, the neocortical layers do not all become affected at the same time, nor are the lesions distributed uniformly across supra- and infragranular layers. Instead, it appears that only those pyramidal cells become affected that send projections to the striatum, thalamic relay nuclei, claustrum, and amygdala. In other words, only a very specific efferent function of the cerebral cortex is lost, but the end result is a substantial reduction of the data stream through the striatal circuit. It is questionable, however, whether this circumscribed or specific type of cortical damage is sufficient to induce, much less explain, the development of a demential syndrome.

The existence and morphological preservation of phylogenetically late appearing and ontogenetically late maturing supragranular layers (IIIa,b) are generally regarded as being necessary for associative capabilities. Inasmuch as these layers specifically remain largely unaffected, the factors that contribute to the emergence of dementia in sPD must be different than those operative in Alzheimer's disease.

Both disorders have features in common, namely, severe inclusion body pathologies and neuronal destruction bilaterally in the anteromedial temporal mesocortex, which in turn interrupt the flow of data through the limbic circuit. The ensuing neuronal damage within the entorhinal region and hippocampal formation is, nevertheless—when compared to that in Alzheimer's disease—much milder in sPD, so that the question is justified whether the cortical Lewy pathology becomes severe enough to cause dementia. A large number of findings show that functions of the basal ganglia are not exclusively confined to movement but fulfill a multiplicity of tasks, including those related to cognition and emotional behavior (Saper 1996; Parent et al. 2000). Symptoms indicative of a demential syndrome can appear very early in sPD, i.e., sometimes prior to the manifestation of the classical motor symptoms (Braak et al. 2005). The issue becomes even more complicated when the Lewy pathology co-occurs—which is frequently the case—with Alzheimer-related neurofibrillary pathology but escapes recognition clinically because currently existing diagnostic tools are not sophisticated enough to differentiate "pure" Alzheimer's cases from cases with both pathologies. Similar to the nonmotor symptoms, dementia in sPD requires further study before it can be adequately understood.

8.6 Pathologically Changed Components of the Cerebellar Circuit

The cerebellar circuit includes the pons, cerebellum, and select lower brainstem nuclei. Components of this circuit conduct necessary information regarding muscle tonus, position, and movement of body parts to ensure the normal functioning of the corticospinal pyramidal tract. Together with the centers of the striatal and limbic circuits, all three circuits enable the organism to execute synergistic, modulated, and purposeful movements (Fig. 29). The thalamus integrates input from the cerebellar circuit and influences cortical areas that generate corticonuclear and corticospinal projections to premotor and motor neurons of the lower brainstem and spinal cord (Fig. 29). The precerebellar nuclei in the lower brainstem receive data from the cerebral cortex either directly (e.g., cortico-olivary tract) or indirectly via relay stations (e.g., corticorubral tract, rubro-olivary tract), whereas ascending input arrives from the spinal cord (e.g., spino-olivary tract). Most precerebellar centers (lateral reticular nucleus, external cuneate nucleus, vestibular nuclei, inferior olivary nucleus, dorsal paramedian reticular nucleus, pontine gray, arcuate nucleus, pontobulbar body) receive both descending and ascending input. All precerebellar nuclei send projections to the cerebellum. The inferior olivary nucleus generates climbing fibers; all other nuclei give rise to mossy fibers. The data transmitted are processed in the cerebellar cortex and conveyed, for the most part, via the dentate nucleus and ventrolateral posterior nucleus of the thalamus (VLp), to neocortical motor areas, which generate descending projections to medullary and spinal cord premotor and motor neurons (Fig. 29).

The cerebellar cortex and the centers of the cerebellar circuit commence myelination prenatally. Their long-axoned projection neurons develop medium to thick-caliber

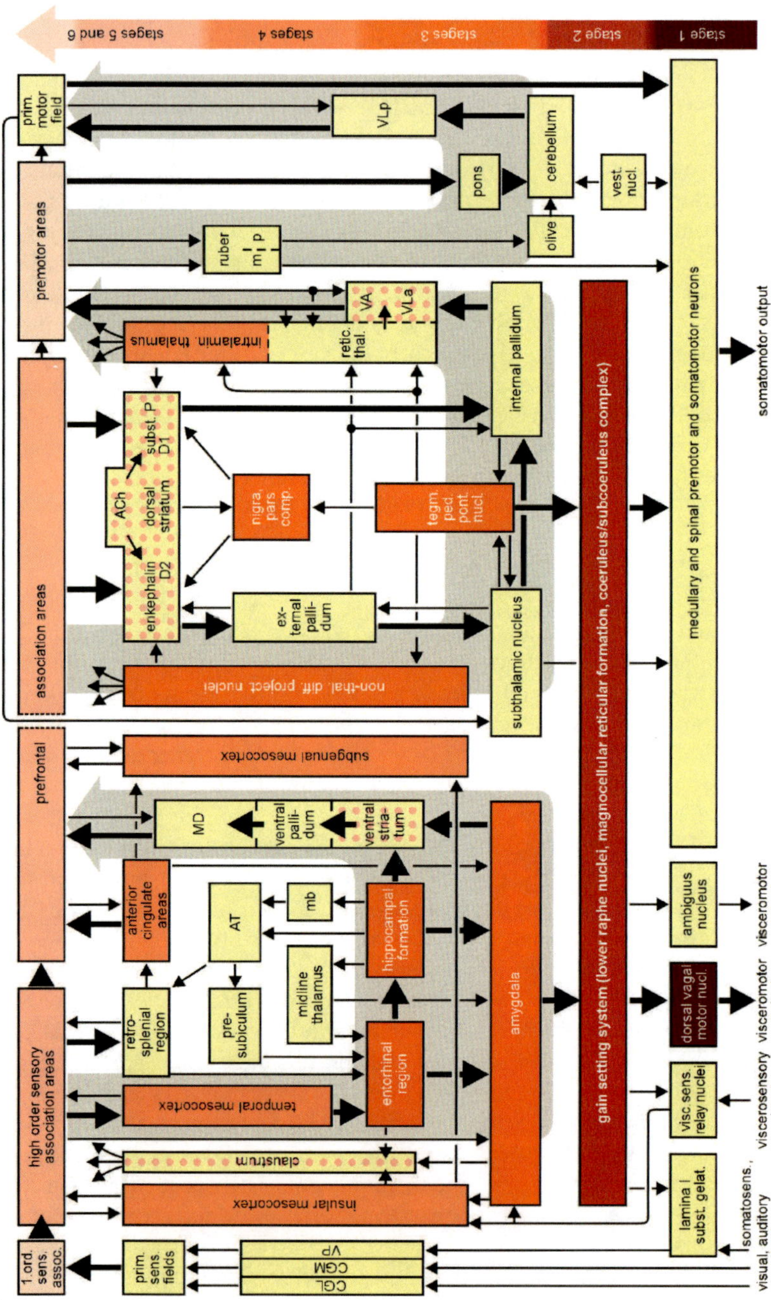
stage 1
stage 2
stages 3
stages 4
stages 5 and 6
prim. motor field
premotor areas
association areas
prefrontal
high order sensory association areas
1.ord. sens. assoc.
prim. sens. fields
CGL
CGM
VP
VLp
pons
cerebellum
vest. nucl.
olive
ruber
m p
VA
VLa
retic. thal.
intralamin. thalamus
internal pallidum
ACh
dorsal striatum
enkephalin D2
subst. P D1
nigra, pars comp.
tegm. ped. pont. nucl.
ex-ternal palli-dum
subthalamic nucleus
non-thal. diff. project. nuclei
subgenual mesocortex
MD
ventral palli-dum
ventral stria-tum
anterior cingulate areas
AT
mb
hippocampal formation
retro-splenial region
pre-subiculum
midline thalamus
entorhinal region
amygdala
temporal mesocortex
claustrum
insular mesocortex
gain setting system (lower raphe nuclei, magnocellular reticular formation, coeruleus/subcoeruleus complex)
medullary and spinal premotor and somatomotor neurons
somatomotor output
ambiguus nucleus
visceromotor output
dorsal vagal motor nucl.
visceromotor output
visc. sens. relay nuclei
viscerosensory input
lamina I subst. gelat.
somatosens., visual, auditory input

Fig. 29 Schema showing the major components of the limbic, visceromotor, and somatomotor systems. Important centers of the cerebellar circuit have been added to the diagram to make the somatomotor system more complete. Additional details are supplied in the form of select cortical areas that regulate viscerosensory and visceromotor functions (insular and subgenual mesocortex). Moving from *left to right*, the three broad *gray arrows* (background) are intended to facilitate recognition of the limbic, striatal, and cerebellar circuits. Even in this much more complicated diagram, it becomes obvious that normal functions of the limbic circuit depend on the structural integrity of the temporal mesocortex. It is characteristic of neuropathological stages 5–6 that the lesions advance into the mature neocortex, initially making inroads into the extended prefrontal and high order sensory association areas (stage 5), followed by incursions into premotor and first order sensory association areas, and eventually the primary fields (stage 6). Affected nuclei or regions are marked by varying degrees of shading: *dark brown* (involved from stage 1 onwards), *dark red* (from stage 2 onwards), *red and light red* (from stages 3 and 4 onwards), *pink* (from stage 5 onwards), and *light pink* (first involved in stage 6). *Dotted-Pink* regions indicate presence of only immunoreactive terminal nerve cell processes and/or astrocytes. *Yellow* indicates either noninvolvement. Centers of the limbic and striatal circuits undergo the most severe neurodegeneration. The severe involvement of the anteromedial temporal mesocortex leads to a pronounced reduction of the data-transfer from the sensory neocortex via the entorhinal region, hippocampal formation, and amygdala to the prefrontal cortex. The gradual involvement of the neocortex can pave the way for cognitive impairment in sporadic Parkinson's disease. **Abbreviations:** *ACh*, Large cholinergic local circuit neurons of the striatum; *AT*, anterior thalamic nuclei; *CGL*, lateral geniculate body; *CGM*, medial geniculate body; *enkephalin*, enkephalin-producing projection neurons of the dorsal striatum; *inf. olive*, principal nucleus of the inferior olive; *intralam. thalamus*, intralaminar nuclei of the thalamus; *mb*, mamillary body; *MD*, mediodorsal nuclei of the thalamus; *m/p ruber*, magnocellular and parvocellular portions of the red nucleus; *prim. sens. fields*, primary sensory fields; *retic. thal.*, reticular nucleus of the thalamus; *somatosens.*, somatosensory; *subst. gelat.*, substantia gelatinosa; *subst. nigra, pars comp.*, substantia nigra, pars compacta; *subst. P*, substance P-producing projection neurons of the dorsal striatum; *tegm ped pont nucl.*, pedunculopontine tegmental nucleus; *VA*, ventral anterior nucleus of the thalamus; *ventr. pallidum*, ventral pallidum; *vest. nucl.*, vestibular nuclei; *visc. sens.*, viscerosensory; *VLa*, anterior ventrolateral nucleus of the thalamus; *VLp*, posterior ventrolateral nucleus of the thalamus; *VP*, ventral posterior nuclear complex of the thalamus; *1. ord. sens. assoc.*, first order sensory association areas

myelin sheaths and, in general, their nerve cells are resistant to Lewy pathology or seldom become involved in sPD—and when they do, it is in the final stages of the disorder. We have occasionally observed slight and patchy involvement of the inferior olivary nucleus in sPD cases at stages 5 and 6 (Fig. 28b,c), but the same cannot be said to date for any of the other precerebellar centers listed above. From the standpoint of the larger pathological process, such an event can be interpreted as the initial foray into what is normally an "off limits" area of the CNS—but it also intimates the further trajectory of the pathological process in sPD, provided individuals with the severe neuronal damage that characterizes end stage cases survive for longer periods of time.

9 The Progression of the Cortical Lesions Mimics the Pattern of Myelination in Reverse Order

The primary fields of the neocortex are the point of departure of neocortical evolution and, as such, they command the most space in the cerebral cortex of lower mammals and early primates. Next come the less expansive and less refined first order sensory association and premotor areas. Least spread out are the most elementarily organized highorder sensory association and prefrontal areas. Further along the primate evolutionary scale, the simply organized high order association areas expand relative to that of the primary fields. With growing distance from the primary fields and increasing proximity to the allocortex, the association areas of the neocortex and anterior mesocortex exhibit greater features of structural immaturity (Braak and Braak 1996).

Myelination represents the final step in brain maturation. Functional maturity of projection neurons usually is achieved only after myelination of the axons is completed. In the human brain, myelination of the neocortex is a late-onset and particularly prolonged process that follows a predetermined sequence (van der Knapp et al. 1991; van der Knaap and Valk 1995; Hasegawa et al. 1992; Nieuwenhuys 1999). It commences in the neocortical primary fields and continues via the first order sensory association areas and premotor fields into high order sensory association areas and prefrontal fields, eventually reaching the mesocortex. Exceptionally dense myelination of the primary fields is the end-result in the human adult and, on the average, myelin density declines with increasing distance from the primary areas. Accordingly, the anteromedial temporal mesocortex is very sparsely myelinated.

Regressive brain changes often tend to repeat the maturation process but in reverse order (Reisberg et al. 1992, 2003; Braak and Braak 1996; Arendt et al. 1998; Cramer and Chopp 2000; Moceri et al. 2000; Braak and Del Tredici 2004; Miller et al. 2008). Just as the primary neocortical fields that are heavily myelinated are more or less impervious to the disease process, cortical areas that myelinate last are the most prone to develop the lesions. As such, it should come as no surprise that the poorly myelinated anteromedial temporal mesocortex is the induction site of the earliest cortical LNs and LBs (Braak et al. 2003a).

10 The Staging Hypothesis: Assumptions, Challenges, Potential

Neuroanatomical studies that take into account morphological features, such as myelination, pigmentation, and other characteristics of susceptible neuronal types, as well as the existence of anatomical pathways that interconnect involved nervous system sites, can help to provide insight into the pathological process that underlies sPD. A solid anatomical and morphological basis is required for accurate clinicopathological correlations.

The concept that the sPD-related pathological process is a progressive one rests on a few basic assumptions: (1) The synucleinopathy sPD is a single neuropathological entity that affects portions of the entire human nervous system. (2) The Lewy pathology in selectively vulnerable nerve cell types (projection neurons with long and poorly myelinated axons) causes functional decline and, eventually, cell loss of the parent neurons at specific sites. (3) Lewy pathology in sPD is age-related but not an inevitable concomitant of the aging process. (4) Incidental Lewy pathology found in nonsymptomatic individuals represents a premorbid state. Lesion density and individual differences in disposition, including existing comorbidities, probably determine when the threshold to the clinically recognizable phase (motor phase) of the disorder is crossed. (5) Physical contacts between projection cells of vulnerable regions may play a key role in the pathogenesis of sPD inasmuch as all of the vulnerable regions within the nervous system are interconnected anatomically.

Large, autopsy-controlled prospective studies that include healthy individuals (controls) are needed to test the reproducibility of the neuropathological staging procedure for sPD (Litvan et al. 2007; Dickson et al. 2008). Some of the earliest available results indicate a degree of accuracy that ranges from 60% (Halliday et al. 2008) to 83% (Parkkinen et al. 2008) or 94% (Duda et al. 2007). In the final analysis, the staging hypothesis has helped to precipitate a shift away from the nearly exclusive focus on the dopaminergic system and the (relatively speaking) late symptomatic features toward a greater awareness of the extranigral pathology accompanied by a heightened interest in early nonmotor symptoms and development of causally oriented therapies (Litvan et al. 2007; Del Tredici and Braak 2008).

Acknowledgements

Funding for this project was made possible, in part, by the German Research Council (Deutsche Forschungsgemeinschaft), the Hilde Ulrichs Foundation (Florstadt-Staden, Germany), and the Michael J. Fox Foundation (United States). We wish to express our thanks to our colleague Prof. Horst-Werner Korf, M.D. (Dr. Senckenberg Anatomical Institute, University of Frankfurt) for inviting us to prepare this monograph, as well as to Mr. M. Bouzrou (immunocytochemistry) and Ms. I. Szász-Jacobi (graphics) for their adept technical support.

References

Aarsland D, Andersen K, Larsen JP, Lolk A, Kragh-Sorensen P (2003) Prevalence and characteristics of dementia in Parkinson disease. An 8-year prospective study. Arch Neurol 60:387–392

Abbott RD, Petrovitch H, White LR, Masaki KH, Tanner CM, Curb JD, Grandinetti A, Blanchette PL, Popper JS, Ross GW (2001) Frequency of bowel movements and the future risk of Parkinson's disease. Neurology 57:456–462

Abbott RD, Ross GW, White LR, Tanner CM, Masaki KH, Nelson JS, Curb JD, Petrovitch H (2005) Excessive daytime sleepiness and subsequent development of Parkinson disease. Neurology 65:1442–1446

Abbott RD, Ross GW, Petrovitch H, Tanner CM, Davis DG, Masaki KH, Launer LJ, Curb JD, White LR (2007) Bowel movement frequency in late-life and incidental Lewy bodies. Mov Disord 22:1581–1586

Adler CH (2005) Nonmotor complications in Parkinson's disease. Mov Disord 20:23–29

Adler CH, Thorpy MJ (2005) Sleep issues in Parkinson's disease. Neurology 64:12–20

Aggleton JP (2000) The amygdala: a functional analysis. University Press, Oxford

Agid Y, Ruberg M, Javoy-Agid F, Hirsch E, Raisman-Vozari R, Vyas S, Faucheux B, Michel P, Kastner A, Blanchard V, Damier P, Villares J, Zhang P (1993) Are dopaminergic neurons selectively vulnerable to Parkinson's disease? Adv Neurol 60:148–164

Ahlskog JE (2005) Challenging conventional wisdom: the etiologic role of dopamine oxidative stress in Parkinson's disease. Mov Disord 20:271–282

Ahlskog JE (2007) Beating a dead horse: dopamine and Parkinson's disease. Neurology 69:1701–1711

Albin RL, Young AB, Peney JB (1995) The functional anatomy of disorders of the basal ganglia. Trends Neurosci 18:63–64

Alegre-Abarrategui J, Ansorge O, Esiri M, Wade-Martins R (2007) LRRK2 is a component of granular alpha-synuclein pathology in the brainstem of Parkinson's disease. Neuropathol Appl Neurobiol 34:272–283

Alexander GE, Crutcher MD, DeLong MR (1990) Basal ganglia-thalamocortical circuits: parallel substrates for motor, oculomotor, "prefrontal" and "limbic" functions. Prog Brain Res 85:119–146

Alheid GF (2003) Extended amygdala and basal forebrain. Ann N Y Acad Sci 985:185–205

Alheid GF, Heimer L, Switzer RC (1990) Basal ganglia. In: Paxinos G (ed) The human nervous system. Academic Press, San Diego, pp 483–582

Alim MA, Ma QL, Takeda K, Aizawa T, Matsuara M, Nakamura M, Asada A, Saito T, Kaji H, Yoshii M, Hisanaga S, Uéda K (2004) Demonstration of a role for α-synuclein as a functional microtubule-associated protein. J Alzheimers Dis 6:435–442

Altschuler E (1996) Gastric helicobacter pylori infection as a cause of idiopathic Parkinson disease and non-arteric anterior optic ischemic neuropathy. Med Hypotheses 47:413–414

Amaral DG, Price JL, Pitkänen A, Carmichael ST (1987) Anatomical organization of the primate amygdaloid complex. In: Aggleton JP (ed) The amygdala: neurobiological aspects of emotion, memory, and mental dysfunction. Wiley-Liss, New York, pp 1–66

Amino T, Orimo S, Itoh Y, Takahashi A, Uchihara T, Mizusawa H (2005) Profound cardiac sympathetic denervation occurs in Parkinson's disease. Brain Pathol 15:29–34

Anderson JP, Walker DE, Goldstein JM, de Laat R, Banducci K, Caccavello RJ, Barbour R, Huang J, Kling K, Lee M, Diep L, Keim PS, Shen X, Chataway T, Schlossmacher MG, Seubert P, Schenk D, Sinha S, Gai WP, Chilcote TJ (2006) Phosphorylation of Ser 129 is the dominant pathological modification of alpha-synuclein in familial and sporadic Lewy body disease. J Biol Chem 281:29739–29752

Anlauf M, Schäfer MKH, Eiden L, Weihe E (2003) Chemical coding of the human gastrointestinal nervous system: cholinergic, VIPergic, and catecholaminergic phenotypes. J Comp Neurol 459:90–111

Arai K, Kato N, Kashiwado K, Hattori T (2000) Pure autonomic failure in association with human alpha-synucleinopathy. Neurosci Lett 296:171–173

Arai T, Uéda K, Ikeda K, Akiyama H, Haga C, Kondo H, Kuroki N, Niizato K, Iritani S, Tsuchiya K (1999) Argyrophilic glial inclusions in the midbrain of patients with Parkinson's disease and diffuse Lewy body disease are immunopositive for NACP/alpha-synuclein. Neurosci Lett 259:83–86

Aravamuthan BR, Muthusamy KA, Stein JF, Aziz TZ, Johansen-Berg H (2007) Topography of cortical and subcortical connections of the human pedunculopontine and subthalamic nuclei. NeuroImage 37:694–703

Arawaka S, Saito Y, Murayama S, Mori H (1998) Lewy body in neurodegeneration with brain iron accumulation type 1 is immunoreactive for α-synuclein. Neurology 51:887–889

Arendt T, Brückner MK, Gertz HJ, Marcova L (1998) Cortical distribution of neurofibrillary tangles in Alzheimer's disease matches the pattern of neurons that retain their capacity of plastic remodelling in the adult brain. Neuroscience 83:991–1002

Arima K, Uéda K, Sunohara N, Hirai S, Izumiyama Y, Tonozuka-Uehara H, Kawai M (1998a) Immunoelectronmicroscopic demonstration of NACP/α-synuclein-epitopes on the filamentous component of Lewy bodies in Parkinson's disease and in dementia with Lewy bodies. Brain Res 808:93–100

Arima K, Uéda K, Sunohara N, Arakawa K, Hirai S, Nakamura M, Tonozuka-Uehara H, Kawai M (1998b) NACP/alpha-synuclein immunoreactivity in fibrillary components of neuronal and oligodendroglial cytoplasmic inclusions in the pontine nuclei in multiple system atrophy. Acta Neuropathol 96:439–444

Attems J, Quass M, Jellinger KA (2007) Tau and α-synuclein brainstem pathology in Alzheimer disease: relation with extrapyramidal signs. Acta Neuropathol 113:53–62

Augustine JR (1996) Circuitry and functional aspects of the insular lobe in primates including humans. Brain Res Brain Res Rev 22:229–244

Baba M, Nakajo S, Tu PH, Tomita T, Nakaya K, Lee VMY, Trojanowski JQ, Iwatsubo T (1998) Aggregation of α-synuclein in Lewy bodies of sporadic Parkinson's disease and dementia with Lewy bodies. Am J Pathol 152:879–884

Baker H, Genter MB (2003) The olfactory system and the nasal mucosa as portals of entry of viruses, drugs, and other exogenous agents into the brain. In: Doty RL (ed) Handbook of olfaction and gustation. Marcel Dekker, New York, pp 549–573

Barbas H, Rempel-Clower N (1997) Cortical structure predicts the pattern of corticocortical connections. Cerebral Cortex 7:635–646

Beach TG, McGeer EG (1984) The distribution of substance P in the primate basal ganglia: an immunohistochemical study of baboon and human brain. Neuroscience 13:29–52

Beach TG, Adler CH, Sue LL, Peirce JB, Bachalakuri J, Dalsing-Hernandez JE, Lue LF, Caviness JN, Connor DJ, Sabbagh M, Walker DG (2008) Reduced striatal tyrosine hydroxylase in incidental Lewy body disease. Acta Neuropathol 115:445–451

Beal MF (1995) Aging, energy, and oxidative stress in neurodegenerative diseases. Ann Neurol 38:357–366

Benabid AL (2003) Deep brain stimulation for Parkinson's disease. Curr Opin Neurobiol 13:696–706

Benarroch EE (1993) The central autonomic network: functional organization, dysfunction, and perspective. Mayo Clin Proc 68:988–1001

Benarroch EE (2006) Pain-autonomic interactions. Neurol Sci 27:130–133

Benarroch EE (2007) Enteric nervous system: functional organization and neurologic implications. Neurology 69:1953–1957

Benarroch EE, Chang FLF (1993) Central autonomic disorders. J Clin Neurophysiol 10:39–50

Benarroch EE, Schmeichel AM, Sandroni P, Low PA, Parisi JE (2006) Involvement of vagal autonomic nuclei in multiple system atrophy and Lewy body diseases. Neurology 66:378–383

Bentivoglio M, Kultas-Ilinksy K, Ilinsky I (1993) Limbic thalamus: structure, intrinsic organization, and connections. In: Vogt BA, Gabriel M (eds) Neurobiology of cingulate cortex and limbic thalamus. Birkhäuser, Boston, pp 71–122

Berendse HW, Groenewegen HJ (1991) Restricted cortical termination fields of the midline and intralaminar thalamic nuclei in the rat. Neuroscience 42:73–102

Berendse HW, Booij J, Francot CMJE, Bergmans PLM, Hijman R, Stoof JC, Wolters EC (2001) Subclinical dopaminergic dysfunction in asymptomatic Parkinson's disease patients' relatives with a decreased sense of smell. Ann Neurol 50:34–41

Bergeron C, Pollanen MS (1996) Pathogenesis of the Lewy body. In: Perry RH, McKeith IG, Perry EK (eds) Dementia with Lewy bodies. Cambridge University Press, New York, pp 302–307

Beró T, Nagy L, Jávor T, Vincze A (1992) Measurement of non-steroid antiinflammatory drugs induced gastric microbleeding. Acta Physiol Hung 80:281–287

Betchen SA, Kaplitt M (2003) Future and current surgical therapies in Parkinson's disease. Curr Opin Neurol 16:487–493

Biere AL, Wood SJ, Wypych J, Steavenson S, Jiang Y, Anafi D, Jacobsen FW, Jarosinski MA, Wu GM, Louis JC, Martin F, Narhi LO, Citron M (2000) Parkinson's disease-associated alpha synuclein is more fibrillogenic than beta- and gamma-synuclein and cannot crossseed its homologs. J Biol Chem 275:34574–34579

Blandini F, Nappi G, Tassorelli C, Martignioni E (2000) Functional changes of the basal ganglia circuitry in Parkinson's disease. Prog Neurobiol 62:63–88

Blessing WW (2004) Lower brain stem regulation of visceral, cardiovascular, and respiratory function. In: Paxinos G, Mai JK (eds) The human nervous system, 2nd edn. Elsevier, San Diego, pp 464–478

Bloch A, Probst A, Bissig H, Adams H, Tolnay M (2006) α-Synuclein pathology of the spinal and peripheral autonomic nervous system in neurologically unimpaired elderly subjects. Neuropathol Appl Neurobiol 12:284–295

Boeve BF, Saper CB (2006) REM sleep behavior disorder: a possible early marker for synucleinopathies. Neurology 28:796–797

Boeve BF, Silber MH, Parisi JE, Dickson DW, Ferman TJ, Benarroch EE, Schmeichel AM, Smith GE, Petersen RC, Ahlskog JE, Matsumoto JY, Knopman DS, Schenck CH, Mahowald MW (2003) Synucleinopathy pathology and REM sleep behavior disorder plus dementia or parkinsonism. Neurology 61:40–45

Boeve BF, Silber MH, Saper CB, Ferman TJ, Dickson DW, Parisi JE, Benarroch EE, Ahlskog JE, Smith GE, Caselli RC, Tippman-Peikert M, Olson EJ, Lin SC, Wszolek Z, Schenck CH,

Mahowald MW, Castillo PR, Del Tredici K, Braak H (2007) Pathophysiology of REM sleep behaviour disorder and relevance to neurodegenerative disease. Brain 129:3103–3114

Bohus B, Koolhaas JM, Luiten PGM, Korte SM, Roozendaal B, Wiersma A (1996) The neurobiology of the central nuleus of the amygdala in relation to neuroendocrine and autonomic outflow. Prog Brain Res 107:447–460

Borghi R, Marchese R, Negro A, Marinelli L, Forloni G, Zaccheo D, Abbruzzese G, Tabaton M (2000) Full length α-synuclein is present in cerebrospinal fluid from Parkinson's disease and normal subjects. Neurosci Lett 287:65–67

Bower JH, Maraganore DM, Peterson BJ, McDonell SK, Ahlskog JE, Rocca WA (2003) Head trauma preceding PD: a case-control study. Neurology 60:1610–1615

Braak E, Sandmann-Keil D, Rüb U, Gai WP, de Vos RAI, Jansen Steur ENH, Arai K, Braak H (2001) Alpha-synuclein immunopositive Parkinson's disease-related inclusion bodies in lower brainstem nuclei. Acta Neuropathol 101:195–201

Braak H (1980) Architectonics of the human telencephalic cortex. Springer, Berlin

Braak H (1984) Architectonics as seen by lipofuscin stains. In: Jones EG, Peters A (eds) Cerebral cortex. Cellular components of the cerebral cortex, vol I. Plenum Press, London, pp 59–104

Braak H, Braak E (1984) Neuronal types in the neocortex-dependent lateral territory of the human thalamus. Anat Embryol 169:61–72

Braak H, Braak E (1985) On areas of transition between entorhinal allocortex and temporal isocortex in the human brain. Normal morphology and lamina-specific pathology in Alzheimer's disease. Acta Neuropathol 68:325–332

Braak H, Braak E (1986) Nuclear configuration and neuronal types of the nucleus niger in the brain of the human adult. Hum Neurobiol 5:71–82

Braak H, Braak E (1992) The human entorhinal cortex: normal morphology and lamina-specific pathology in various diseases. Neurosci Res 15:6–31

Braak H, Braak E (1996) Development of Alzheimer-related neurofibrillary changes in the neocortex inversely recapitulates cortical myelogenesis. Acta Neuropathol 92:197–201

Braak H, Del Tredici K (2004) Poor and protracted myelination as a contributary factor to neurodegenerative disorders. Neurobiol Aging 25:19–23

Braak H, Del Tredici K (2005) Preclinical and clinical stages of intracerebral inclusion body pathology in idiopathic Parkinson's disease. In: Willow JM (ed) Parkinson's disease: progress in research. Nova Science, New York, pp 1–49

Braak H, Del Tredici K (2008) Nervous system pathology in sporadic Parkinson's disease. Neurology 70:1916–1925

Braak H, Braak E, Yilmazer D, de Vos RAI, Jansen ENH, Bohl J, Jellinger K (1994) Amygdala pathology in Parkinson's disease. Acta Neuropathol 88:493–500

Braak H, Braak E, Yilmazer D, Bohl J (1996) Functional anatomy of human hippocampal formation and related structures. J Child Neurol 11:265–275

Braak H, de Vos RAI, Jansen ENH, Bratzke H, Braak E (1998) Neuropathological hallmarks of Alzheimer's and Parkinson's diseases. Prog Brain Res 117:267–285

Braak H, Sandmann-Keil D, Gai WP, Braak E (1999) Extensive axonal Lewy neurites in Parkinson's disease: a novel pathological feature revealed by α-synuclein immunocytochemistry. Neurosci Lett 265:67–69

Braak H, Rüb U, Sandmann-Keil D, Gai WP, de Vos RAI, Jansen Steur ENH, Arai K, Braak E (2000) Parkinson's disease: affection of brain stem nuclei controlling premotor and motor neurons of the somatomotor system. Acta Neuropathol 99:489–495

Braak H, Del Tredici K, Gai WP, Braak E (2001) Alpha-synuclein is not a requisite component of synaptic boutons in the adult human central nervous system. J Chem Neuroanat 20:245–252

Braak H, Del Tredici K, Rüb U, de Vos RAI, Jansen Steur ENH, Braak E (2003a) Staging of brain pathology related to sporadic Parkinson's disease. Neurobiol Aging 24:197–210

Braak H, Rüb U, Gai WP, Del Tredici K (2003b) Idiopathic Parkinson's disease: possible routes by which vulnerable neuronal types may be subject to neuroinvasion by an unknown pathogen. J Neural Transm 110:517–536

Braak H, Rüb U, Del Tredici K (2003c) Involvement of precerebellar nuclei in multiple system atrophy. Neuropathol Appl Neurobiol 29:60–76

Braak H, Ghebremedhin E, Rüb U, Bratzke H, Del Tredici K (2004) Stages in the development of Parkinson's disease-related pathology. Cell Tissue Res 318:121–134

Braak H, Rüb U, Jansen Steur ENH, Del Tredici K, de Vos RAI (2005) Cognitive status correlates with neuropathological stage in Parkinson disease. Neurology 64:1404–1410

Braak H, de Vos RAI, Bohl J, Del Tredici K (2006a) Gastric α-synuclein immunoreactive inclusions in Meissner's and Auerbach's plexuses in cases staged for Parkinson's disease-related brain pathology. Neurosci Lett 396:67–72

Braak H, Müller CM, Rüb U, Ackermann H, Bratzke H, de Vos RAI, Del Tredici K (2006b) Pathology associated with sporadic Parkinson's disease—where does it end? J Neural Transm 70:89–97

Braak H, Müller CM, Bohl JR, Rüb U, de Vos RAI, Del Tredici K (2006c) Stanley Fahn Lecture 2005: the staging procedure for the inclusion body pathology associated with sporadic Parkinson disease reconsidered. Mov Disord 21:2042–2051

Braak H, Sastre M, Bohl JRE, de Vos RAI, Del Tredici K (2007a) Parkinson's disease: lesions in dorsal horn layer I, involvement of parasympathetic and sympathetic pre- and post-ganglionic neurons. Acta Neuropathol 113:421–429

Braak H, Sastre M, Del Tredici K (2007b) Development of α-synuclein immunoreactive astrocytes in the forebrain parallels stages of intraneuronal pathology in sporadic Parkinson's disease. Acta Neuropathol 114:231–241

Brady ST, Witt AS, Kirkpatrick LL, de Waegh SM, Readhead C, Tu PH, Lee VMY (1999) Formation of compact myelin is required for maturation of the axonal cytoskeleton. J Neurosci 19:7278–7288

Broadwell RD, Brightman MV (1976) Entry of peroxidase into neurons of the central and peripheral nervous systems from extracerebral and cerebral blood. J Comp Neurol 166:257–284

Brooks DJ (1998) The early diagnosis of Parkinson's disease. Ann Neurol 44:10–18

Brooks DJ (2000) PET studies and motor complications in Parkinson's disease. Trends Neurosci 23:101–108

Brooks DJ, Frey KA, Marek KL, Oakes D, Paty D, Prentice R, Shults CW, Stoessl AJ (2003) Assessment of neuroimaging techniques as biomarkers of the progression of Parkinson's disease. Exp Neurol 184:68–79

Buzas B, Max MB (2004) Pain in Parkinson disease. Neurology 62:2156–2157

Cabello CR, Thune JJ, Pakkenberg H, Pakkenberg B (2002) Ageing of substantia nigra in humans: cell loss may be compensated by hypertrophy. Neuropathol Appl Neurobiol 28:283–291

Calabresi P, Centonze D, Gubellini P, Pisani A, Bernardi G (2000) Acetylcholine-mediated modulation of striatal function. Trends Neurosci 23:120–126

Calne DB, Snow BJ, Lee C (1992) Criteria for diagnosing Parkinson's disease. Ann Neurol 32:125–127

Campbell SK, Switzer RC, Martin TL (1987) Alzheimer's plaques and tangles: a controlled and enhanced silver-staining method. Soc Neurosci Abstr 13:678

Card JP (1998) Exploring brain circuitry with neurotropic viruses: new horizons in neuroanatomy. Anat Rec 253:176–185

Casado B, Pannell LK, Iadarola P, Baraniuk JN (2005) Identification of human nasal mucous proteins using proteomics. Proteomics 5:2949–2959

Cassarino DS, Quezado MM, Ghatak NR, Duray PH (2003) Lyme-associated parkinsonism: a neuropathologic case study and review of the literature. Arch Pathol Lab Med 127:1204–1206

Cechetto DF, Saper CB (1990) Role of the cerebral cortex in autonomic function. In: Loewy AD, Spyer KM (eds) Central regulation of autonomic function. Oxford University Press, New York, pp 208–223

Cersósimo MG, Benarroch EE (2008) Neural control of the gastrointestinal tract: implications for Parkinson disease. Mov Disord (in press)

Chade AR, Kasten M, Tanner CM (2006) Nongenetic causes of Parkinson's disease. J Neural Transm 70:147–151

Chaudhuri KR (2001) Autonomic dysfunction in movement disorders. Curr Opin Neurol 14:505–511

Chaudhuri KR, Healy DG, Schapira AH (2006) Non-motor symptoms of Parkinson's disease: diagnosis and management. Lancet Neurol 5:235–245

Chesselet MF, Delfs JM (1996) Basal ganglia and movement disorders: an update. Trends Neurosci 19:417–422

Chu CC, Tranel D, Damasio AR, van Hoesen GW (1997) The autonomic-related cortex: pathology in Alzheimer's disease. Cerebral Cortex 7:86–95

Chu Y, Kordower JH (2007) Age-associated increases of alpha-synuclein in monkeys and humans are associated with nigrostriatal dopamine depletion: is this the target for Parkinson's disease? Neurobiol Dis 25:134–149

Chua CE, Tang BL (2006) α-Synuclein and Parkinson's disease: the first roadblock. J Cell Mol Med 10:837–846

Chung KKK, Dawson VL, Dawson TM (2001) The role of the ubiquitin-proteasomal pathway in Parkinson's disease and other neurodegenerative disorders. Trends Neurosci 24:7–14

Ciechanover A, Brundin P (2003) The ubiquitin proteasome system in neurodegenerative diseases: sometimes the chicken, sometimes the egg. Neuron 40:427–446

Cissé S, Perry G, Lacoste-Royal G, Cabana T, Gauvreau D (1993) Immunochemical identification of ubiquitin and heat-shock proteins in corpora amylacea from normal aged and Alzheimer's disease brains. Acta Neuropathol 85:233–240

Clayton DF, George JM (1998) The synucleins: a family of proteins involved in synaptic function, plasticity, neurodegeneration and disease. Trends Neurosci 21:249–254

Clayton DF, George JM (1999) Synucleins in synaptic plasticity and neurodegenerative disorders. J Neurosci Res 58:120–129

Cole NB, Murphy DD (2002) The cell biology of alpha-synuclein: a sticky problem? Neuromolecular Med 1:95–109

Costa M, Brookes SJH, Hennig GW (2000) Anatomy and physiology of the enteric nervous system. Gut 47:15–19

Craig AD (1996) An ascending general homeostatic afferent pathway originating in lamina I. Prog Brain Res 107:225–242

Craig AD (2003) Pain mechanisms: labeled lines versus convergence in central processing. Annu Rev Neurosci 26:1–30

Cramer SC, Chopp M (2000) Recovery recapitulates ontogeny. Trends Neurosci 23:265–271

Dale GE, Probst A, Luthert P, Martin J, Anderton BH, Leigh PN (1992) Relationships between Lewy bodies and pale bodies in Parkinson's disease. Acta Neuropathol 83:525–529

Damasio AR (1998) Emotion in the perspective of an integrated nervous system. Brain Res Brain Res Rev 26:83–86

Damasio AR, Tranel D, Damasio H (1990) Individuals with sociopathic behavior caused by frontal damage fail to respond automatically to social stimuli. Behav Brain Res 41:81–94

Damier P, Hirsch EC, Agid Y, Graybiel AM (1999a) The substantia nigra of the human brain. I. Nigrosomes and the nigral matrix, a compartmental organization based on calbindin D28K immunohistochemistry. Brain 122:1421–1436

Damier P, Hirsch EC, Agid Y, Graybiel AM (1999b) The substantia nigra of the human brain. II. Patterns of loss of dopamine-containing neurons in Parkinson's disease. Brain 122:1437–1448

Daniel SE, Hawkes CH (1992) Preliminary diagnosis of Parkinson's disease by olfactory bulb pathology. Lancet 340:186

Day M, Wang Z, Ding J, An X, Ingham C, Shering AF, Wokosin D, Ilijic E, Sun Z, Sampson AR, Mugnaini E, Deutch AY, Sesack SR, Arbuthnott GW, Surmeier DJ (2006) Selective elimination of glutamatergic synapses on striatopallidal neurons in Parkinson disease models. Nat Neurosci 9:251–259

De Lau L, Breteler MM (2006) Epidemiology of Parkinson's disease. Lancet Neurol 5:525–535

de Vos RAI, Jansen ENH, Stam FC, Ravid R, Swaab D (1995) Lewy body disease: clinico-pathological correlations in 18 consecutive cases of Parkinson's disease with and without dementia. Clin Neurol Neurosurg 97:13–22

de Vos RAI, Jansen ENH, Yilmazer D, Braak H, Braak E (1996) Pathological and clinical features of Parkinson's disease with and without dementia. In: Perry RH, McKeith IG, Perry EK (eds) Dementia with Lewy bodies. Cambridge University Press, New York, pp 255–267

Del Tredici K, Braak H (2004) Idiopathic Parkinson's disease: staging an α-synucleinopathy with a predictable pathoanatomy. In: Kahle P, Haass C (eds) Molecular mechanisms in parkinson's disease. Landes Bioscience, Georgetown, pp 1–32

Del Tredici K, Braak H (2008) A not entirely benign procedure: progression of Parkinson's disease. Acta Neuropathol 115:379–384

Del Tredici K, Rüb U, de Vos RAI, Bohl JRE, Braak H (2002) Where does Parkinson disease pathology begin in the brain? J Neuropathol Exp Neurol 61:413–426

DeLong MR (1990) Primate models of movement disorders of basal ganglia. Trends Neurosci 13:281–285

DeLong MR, Wichmann T (2007) Circuits and circuit disorders of the basal ganglia. Arch Neurol 64:20–24

den Hartog Jager WA, Bethlem J (1960) The distribution of Lewy bodies in the central and autonomic nervous systems in idiopathic paralysis agitans. J Neurol Neurosurg Psychiatry 23:283–290

Deutch AY (2006) Striatal plasticity in parkinsonism: dystrophic changes in medium spiny neurons and progression in Parkinson's disease. J Neural Transm 70:67–70

Dickson DW (1999) Tau and synuclein and their role in neuropathology. Brain Pathol 9:657–661

Dickson DW (2001) α-Synuclein and the Lewy body disorders. Curr Opin Neurol 14: 423–432

Dickson DW, Litvan I (2003) Corticobasal degeneration. In: Dickson DW (ed) Neurodegeneration: the molecular pathology of dementia and movement disorders. ISN Neuropathologica Press, Basel, pp 115–123

Dickson DW, Ruan D, Crystal H, Mark MH, Davies P, Kress Y, Yen SH (1991) Hippocampal degeneration differentiates diffuse Lewy body disease (DLBD) from Alzheimer's disease: light and electron microscopic immunocytochemistry of CA2-3 neurites specific to DLBD. Neurology 41:1402–1409

Dickson DW, Schmidt ML, Lee VMY, Zhao ML, Yen SH, Trojanowski JQ (1994) Immunoreactivity profile of hippocampal CA2/3 neurites in diffuse Lewy body disease. Acta Neuropathol 87:269–276

Dickson DW, Uchikado H, Klos KJ, Josephs KA, Boeve BF, Ahlskog JE (2006) A critical review of the Braak staging scheme for Parkinson's disease. Mov Disord 21:S559(P846)

Dickson DW, Rademakers R, Hutton ML (2007) Progressive supranuclear palsy: pathology and genetics. Brain Pathol 17:74–82

Dickson DW, Fujishiro H, DelleDonne A, Menke J, Zeshan A, Klos K, Josephs K, Frigerio R, Burnett M, Parisi J, Ahlskog JE (2008) Evidence that incidental Lewy body disease is pre-symptomatic Parkinson's disease. Acta Neuropathol 115:437–444

Ding Q, Keller JN (2001) Proteasomes and proteasome inhibition in the central nervous system. Free Radic Biol Med 31:574–584

Djaldetti R, Shifrin A, Rogowski Z, Sprecher E, Melamed E, Yarnitsky D (2004) Quantitative measurement of pain sensation in patients with Parkinson disease. Neurology 62:2171–2175

Dobbs SM, Dobbs RJ, Weller C, Charlett A (2000) Link between Helicobacter pylori infection and idiopathic parkinsonism. Med Hypotheses 55:93–98

Doty RL (2001) Olfaction. Annu Rev Psychol 52:423–452

Doty RL (2008) The olfactory vector hypothesis of neurodegenerative disease: is it viable? Ann Neurol 63:7–15

Doty RL, Stern MB, Pfeiffer C, Gollomp SM, Hurtig HI (1992) Bilateral olfactory dysfunction in early stage treated and untreated idiopathic Parkinson's disease. J Neurol Neurosurg Psychiatry 55:138–142

Drevets WC, Price JL, Simpson JR, Todd RD, Reich T, Vannier M, Raichl ME (1997) Subgenual prefrontal cortex abnormalities in mood disorders. Nature 386:824–827

Dubois B, Pillon B (1997) Cognitive deficits in Parkinson's disease. J Neurol 244:2–8

Duda JE (2004) Pathology and neurotransmitter abnormalities of dementia with Lewy bodies. Dement Geriatr Cogn Disord 17:3–14

Duda JE, Shah U, Arnold SE, Lee VM, Trojanowski JQ (1999) The expression of alpha-, beta-, and gamma-synucleins in olfactory mucosa. Exp Neurol 160:515–522

Duda JE, Lee VMY, Trojanowski JQ (2000) Neuropathology of synuclein aggregates: new insights into mechanism of neurodegenerative diseases. J Neurosci Res 61:121–127

Duda JE, Giasson BI, Mabon ME, Lee VMY, Trojanowski JQ (2002) Novel antibodies to synuclein show abundant striatal pathology in Lewy body disease. Ann Neurol 52:205–210

Duda JE, Noorigian JV, Petrovitch H, White LR, Ross GW (2007) Pattern of Lewy pathology progression suggested by Braak staging system is supported by analysis of a population-based cohort of patients. The Movement Disorders Society's 11th International Congress, Late Breaking Abstracts A3

Duyckaerts C (2005) Atypical Parkinsonian disorders: neuropathology and nosology. In: Litvan I (ed) Current clinical neurology: atypical parkinsonian disorders. Humana Press, Totowa, pp 111–138

Duyckaerts C, Gaspar P, Costa C, Bonnet AM, Hauw JJ (1993) Dementia in Parkinson's disease. Morphometric data. Adv Neurol 60:447–455

El-Agnaf OM, Salem SA, Paleologou KE, Cooper LJ, Fullwood NJ, Gibson MJ, Curran MD, Court JA, Mann DM, Ikeda S, Cookson MR, Hardy J, Allsop D (2003) α-Synuclein implicated in Parkinson's disease is present in extracellular biological fluids, including human plasma. FASEB J 17:1945–1947

El-Agnaf OM, Salem SA, Paleologou KE, Curran MD, Gibson MJ, Court JA, Schlossmacher MG, Allsop D (2005) Detection of oligomeric forms of α-synuclein protein in human plasma as a potential biomarker for Parkinson's disease. FASEB J 20:419–425

Emre M (2004) Dementia in Parkinson's disease: cause and treatment. Curr Opin Neurol 17:399–404

Eriksen JL, Dawson TM, Dickson DW, Petrucelli L (2003) Caught in the act: α-synuclein is the culprit in Parkinson's disease. Neuron 40:453–456

Eriksen JL, Zbigniew W, Petrucelli L (2005) Molecular pathogenesis of Parkinson disease. Arch Neurol 62:353–357

Fahn S (2003) Description of Parkinson's disease as a clinical syndrome. Ann NY Acad Sci 991:1–14

Fallon JH, Loughlin SE (1987) Monoamine innervation of cerebral cortex and a theory of the role of monoamines in cerebral cortex and basal ganglia. In: Jones EG, Peters A (eds) Cerebral cortex. Further aspects of cortical function, including hippocampus, vol 6. Plenum, London, pp 41–127

Fearnley J, Lees A (1994) Pathology of Parkinson's disease. In: Calne DB (ed) Neurodegenerative diseases. Saunders, Philadelphia, pp 545–554

Fearnley JM, Lees AJ (1991) Ageing and Parkinson's disease: substantia nigra regional selectivity. Brain 114:2283–2301

Foley P, Riederer P (1999) Pathogenesis and preclinical course of Parkinson's disease. J Neural Transm 56:31–74

Ford B (1998) Pain in Parkinson's disease. Clin Neurosci 5:63–72

Forno LS (1969) Concentric hyalin intraneuronal inclusions of Lewy type in the brain of elderly persons (50 incidental cases): relationship to parkinsonism. J Am Geriatr Soc 17:557–575

Forno LS (1996) Neuropathology of Parkinson's disease. J Neuropathol Exp Neurol 55:259–272

Forno LS, Langston JW (1993) Lewy bodies and aging: relation to Alzheimer's and Parkinson's diseases. Neurodegeneration 2:19–24

Forno LS, Norville RL (1976) Ultrastructure of Lewy bodies in the stellate ganglion. Acta Neuropathol 34:183–197

Forno LS, DeLanney LE, Irwin I, Langston JW (1993) Similarities and differences between MPTP-induced parkinsonism and Parkinson's disease. Neuropathologic considerations. Adv Neurol 60:600–608

Francois C, Yelnik J, Percheron G (1987) Golgi study of the primate substantia nigra. II. Spatial organization of dendritic arborizations in relation to the cytoarchitectonic boundaries and to the striatonigral bundle. J Comp Neurol 265:473–493

Fudge JL, Haber SN (2000) The central nucleus of the amygdala projection to dopamine subpopulations in primates. Neuroscience 97:479–494

Fujishiro H, Frigerior R, Burnett M, Klos KJ, Josephs JA, DelleDonne A, Parisi JE, Ahlskog JE, Dickson DW (2008) Cardiac sympathetic denervation correlates with clinical and pathologic stages of Parkinson's disease. Mov Disord 23 (in press)

Fujiwara H, Hasegawa M, Dohmae N, Kawashima A, Masliah E, Goldberg MS, Shen J, Takio K, Iwatsubo T (2002) α-Synuclein is phosphorylated in synucleinopathy lesions. Nat Cell Biol 4:160–164

Fumimura Y, Ikemura M, Saito Y, Sengoku R, Kanemaru K, Sawabe M, Arai T, Ito G, Iwatsubo T, Fukayama M, Mizusawa H, Murayama S (2007) Analysis of the adrenal gland is useful for evaluating pathology of the peripheral autonomic nervous system in Lewy body disease. Neuropathol Exp Neurol 66:354–362

Furness JB (2000) Types of neurons in the enteric nervous system. J Auton Nerv Syst 81:87–96

Gai WP, Blumbergs PC, Geffen LB, Blessing WW (1992) Age-related loss of dorsal vagal neurons in Parkinson's disease. Neurology 42:2106–2111

Gai WP, Blessing WW, Blumbergs PC (1995) Ubiquitin-positive degenerating neurites in the brainstem in Parkinson's disease. Brain 118:1447–1459

Gai WP, Power JHT, Blumbergs PC, Blessing WW (1998) Multiple-system atrophy: a new α-synuclein disease? Lancet 352:547–548

Galvin JE, Giasson B, Hurtig HI, Lee VMY, Trojanowski JQ (2000) Neurodegeneration with brain iron accumulations, type 1 is characterized by α-, β-, and γ-synuclein neuropathology. Am J Pathol 157:361–368

Galvin JE, Lee VMY, Trojanowski JQ (2001) Synucleinopathies. Clinical and pathological implications. Arch Neurol 58:186–190

Galvin JE, Pollack J, Morris JC (2006) Clinical phenotype of Parkinson disease dementia. Neurology 67:1605–1611

Garcia-Rill E (1991) The pedunculopontine nucleus. Prog Neurobiol 36:363–389

Gaspar P, Duyckaerts C, Alvarez C, Javoy-Agid F, Berger B (1991) Alterations of dopaminergic and noradrenergic innervations in motor cortex in Parkinson's disease. Ann Neurol 30:365–374

Geddes JW (2005) α-Synuclein: a potent inducer of tau pathology. Exp Neurol 192:244–250

Gelb DJ, Oliver E, Gilman S (1999) Diagnostic criteria for Parkinson's disease. Arch Neurol 56:33–39

Gerfen CR (1992) The neostriatal mosaic: multiple levels of compartmental organization. Trends Neurosci 15:133–139

Gerfen CR (2000) Molecular effects of dopamine on striatal-projection pathways. Trends Neurosci 23:64–70

Gerfen CR (2006) Indirect-pathway neurons lose their spines in Parkinson's disease. Nat Neurosci 9:157–158

Geser F, Wenning GK, Poewe W, McKeith I (2005) How to diagnose dementia with Lewy bodies: state of the art. Mov Disord 20:11–20

Giasson BI, Galvin JE, Lee VMY, Trojanowski JQ (2000a) The cellular and molecular pathology of Parkinson's disease. In: Clark CM, Trojanowski JQ (eds) Neurodegenerative dementias: clinical features and pathological mechanisms. McGraw Hill, New York, pp 219–228

Giasson BI, Jakes R, Goedert M, Duda JE, Leight S, Trojanowski JQ, Lee VM (2000b) A panel of epitope-specific antibodies detects alpha-synuclein in Lewy bodies of Parkinson's disease. J Neurosci Res 59:528–533

Giasson BI, Murray J, Trojanowski JQ, Lee VMY (2001) A hydrophobic stretch of 12 amino acid residues in the middle of α-synuclein is essential for filament assembly. J Biol Chem 276:2380–2386

Giasson BI, Covy JP, Bonini NM, Hurtig HI, Farrer MJ, Trojanowski JQ, Van Deerlin VM (2006) Biochemical and pathological characterization of Lrrk2. Ann Neurol 59:315–322

Gibb WRG (1991) Neuropathology of the substantia nigra. Eur Neurol 31:48–59

Gibb WRG, Lees AJ (1988) The relevance of the Lewy body to the pathogenesis of idiopathic Parkinson's disease. J Neurol Neurosurg Psychiatry 51:745–752

Gibb WRG, Lees AJ (1989) The significance of the Lewy body in the diagnosis of idiopathic Parkinson's disease. Neuropathol Appl Neurobiol 15:27–44

Gilja OH, Hatlebakk JG, Odegaard S, Berstad A, Viola I, Giertsen C, Hausken T, Gregersen H (2007) Advanced imaging and visualization in gastrointestinal disorders. World J Gastroenterol 13:1408–1421

Gilman S (2006) Parkinsonian syndromes. Clin Geriatr Med 22:827–842

Gitler AD, Shorter J (2007) Prime time for alpha-synuclein. J Neurosci 27:2433–2434

Glabe CG, Kayed R (2006) Common structure and toxic function of amyloid oligomers implies a common mechanism of pathogenesis. Neurology 66:74–78

Gloor P (1997) The temporal lobe and limbic system. Oxford University Press, New York

Goedert M (2001) The significance of tau and α-synuclein inclusions in neurodegenerative diseases. Curr Opin Genet Dev 11:343–351

Goedert M, Spillantini MG, Serpell LC, Berriman J, Smith MJ, Jakes R, Crowther RA (2001) From genetics to pathology: tau and α-synuclein assemblies in neurodegenerative diseases. Philos Trans R Soc Lond B Biol Sci 356:213–217

Goetz CG, Luthe W, Tanner CM (1986) Autonomic dysfunction in Parkinson's disease. Neurology 36:73–75

Goetze O, Wieczorek J, Müller T, Przuntek H, Schmidt WE, Woitalla D (2005) Impaired gastric emptying of a solid test meal in patients with Parkinson's disease using 13C-sodium octanoate breadth test. Neurosci Lett 375:170–173

Golbe LI (1999) Alpha-synuclein and Parkinson's disease. Mov Disord 14:6–9

Golbe LI, Mouradian MM (2004) Alpha-synuclein in Parkinson's disease: light from two new angles. Ann Neurol 55:153–156

Goldberg MS, Lansbury PT (2000) Is there a cause-and-effect relationship between α-synuclein fibrillization and Parkinson's disease? Nat Cell Biol 2:E115–E119

Goldman SM, Tanner CM, Oakes D, Bhudhikanok GS, Gupta A, Langston JW (2006) Head injury and Parkinson's disease risk in twins. Ann Neurol 60:65–72

Goldstein DS (2006) Orthostatic hypotension as an early finding in Parkinson's disease. Clin Auton Res 16:46–54

Goldstein DS, Holmes C, Li ST, Bruce S, Metman LV, Cannon RO (2000) Cardiac sympathetic denervation in Parkinson's disease. Ann Intern Med 133:338–347

Gray DA, Woulfe J (2005) Lipofuscin and aging: a matter of toxic waste. Sci Aging Knowledge Environ 5:1–6

Graybiel AM, Hirsch EC, Agid Y (1990) The nigrostriatal system in Parkinson's disease. Adv Neurol 53:17–29

Greffard S, Verny M, Bonnet AM, Beinis JY, Gallinari C, Meaume S, Piette F, Hauw JJ, Duyckaerts C (2006) Motor score of the unified Parkinson disease rating scale as a good predictor of Lewy body-associated neuronal loss in the substantia nigra. Arch Neurol 63:584–588

Gregersen N (2006) Protein misfolding disorders: pathogenesis and intervention. J Inherit Metab Dis 29:456–470

Groenewegen HJ, Berendse HW (1994) The specificity of the 'nonspecific' midline and intralaminar thalamic nuclei. Trends Neurosci 17:52–57

Groenewegen HJ, Trimble M (2007) The ventral striatum as an interface between the limbic and motor systems. CNS Spectr 12:887–892

Grudzien A, Shaw P, Weintraub S, Bigio E, Mash DC, Mesulam MM (2007) Locus coeruleus neurofibrillary degeneration in aging, mild cognitive impairment and early Alzheimer's disease. Neurobiol Aging 28:327–335

Grundy D, Schemann M (2007) Enteric nervous system. Curr Opin Gastroenterol 23:121–126

Haber SN, Gdowski MJ (2004) The basal ganglia. In: Paxinos G, Mai JK (eds) The human nervous system, 2nd edn. Elsevier, London, pp 677–738

Haber SN, Kunishio K, Mizobuchi M, Lynd-Balta E (1995) The orbital and medial prefrontal circuit through the primate basal ganglia. J Neurosci 15:4851–4867

Haehner A, Hummel T, Hummel C, Sommer U, Junghanns S, Reichmann H (2007) Olfactory loss may be a first sign of idiopathic Parkinson's disease. Mov Disord 30:839–842

Hague K, Lento P, Morgello S, Caro S, Kaufmann H (1997) The distribution of Lewy bodies in pure autonomic failure: autopsy findings and review of the literature. Acta Neuropathol 94:192–196

Halliday G, Hely M, Reid W, Morris J (2008) The progression of pathology in longitudinally-followed patients with Parkinson's disease. Acta Neuropathol 115:409–415

Halliday GM (2004) Substantia nigra and locus coeruleus. In: Paxinos G, Mai JK (eds) The human nervous system, 2nd edn. Elsevier, London, pp 449–463

Halliday GM, McCann H (2008) Human-based studies on alpha-synuclein deposition and relationship to Parkinson's disease symptoms. Exp Neurol 209:12–21

Halliday GM, Del Tredici K, Braak H (2006) Critical appraisal of the Braak staging of brain pathology related to sporadic Parkinson's disease. J Neural Transm 70:99–103

Hamani C, Saint-Cyr JA, Fraser J, Kaplitt M, Lozano AM (2004) The subthalamic nucleus in the context of movement disorders. Brain 127:4–20

Hamilton RL (2000) Lewy bodies in Alzheimer's disease: a neuropathological review of 145 cases using α-synuclein immunohistochemistry. Brain Pathol 10:378–384

Hansen LA (1997) The Lewy body variant of Alzheimer disease. J Neural Transm 51:83–93

Hansen LA, Galasko D (1992) Lewy body disease. Curr Opin Neurol Neurosurg 5:889–894

Harding AJ, Stimson E, Henderson JM, Halliday GM (2002) Clinical correlates of selective pathology in the amygdala of patients with Parkinson's disease. Brain 125:2431–2445

Hardoff R, Sula M, Tamir A, Soil A, Front A, Badama S, Honigman S, Giladi N (2001) Gastric emptying time and gastric motility in patients with Parkinson's disease. Mov Disord 16:1041–1047

Hasegawa M, Houdou S, Mito T, Takashima S, Asanuma K, Ohno T (1992) Development of myelination in the human fetal and infant cerebrum: a myelin basic protein immunohistochemical study. Brain Dev 14:1–6

Hauw JJ, Agid Y (2003) Progressive supranuclear palsy (PSP) or Steele-Richardson-Olszewski disease. In: Dickson DW (ed) Neurodegeneration: the molecular pathology of dementia and movement disorders. ISN Neuropathologica Press, Basel, pp 103–114

Hauw JJ, Verny M, Ruberg M, Duyckaerts C (1998) Progressive supranuclear palsy. In: Markesbery WR (ed) Neuropathology of dementing disorders. Arnold, New York, pp 193–218

Hawkes CH (2003) Olfaction in neurodegenerative disorder. Mov Disord 18:364–372

Hawkes CH (2008) Parkinson's disease and aging: same or different process? Mov Disord 23:47–53

Hawkes CH, Shephard BC, Daniel SE (1997) Olfactory dysfunction in Parkinson's disease. J Neurol Neurosurg Psychiatry 62:436–446

Hawkes CH, Shephard BC, Daniel SE (1999) Is Parkinson's disease a primary olfactory disorder? Q J Med 92:473–480

Hawkes CH, Del Tredici K, Braak H (2007) Parkinson's disease: a dual-hit hypothesis. Neuropathol Appl Neurobiol 33:599–614

Hayashida K, Oyanagi S, Mizutani Y, Yokochi M (1993) An early cytoplasmic change before Lewy body maturation: an ultrastructural study of the substantia nigra from an autopsy case of juvenile parkinsonism. Acta Neuropathol 85:445–448

Hazrati LN, Parent A (1991) Projections from the external pallidum to the reticular thalamic nucleus in the squirrel monkey. Brain Res 550:142–146

Heimer L, van Hoesen GW (2006) The limbic lobe and its output channels: implications for emotional functions and adaptive behavior. Neurosci Biobehav Rev 30:126–147

Heimer L, Switzer RD, van Hoesen GW (1982) Ventral striatum and ventral pallidum. Components of the motor system? Trends Neurosci 5:83–87

Heimer L, de Olmos J, Alheid GF, Zaborszky L (1991) "Perestroika" in the basal forebrain: opening the border between neurology and psychiatry. Prog Brain Res 87:109–165

Helke CJ, Adryan KM, Fedorowicz J, Zhuo H, Park JS, Curtis R, Radley HE, Distefano PS (1998) Axonal transport of neurotrophins by visceral afferent and efferent neurons of the vagus nerve of the rat. J Comp Neurol 393:102–117

Hellenbrand W, Seidler A, Robra BP, Vieregge P, Oertel WH, Joerg J, Nischan P, Schneider E, Ulm G (1997) Smoking and Parkinson's disease: a case-control study in Germany. Int J Epidemiol 26:328–339

Henderson JM, Carpenter K, Cartwright H, Halliday GH (2000) Loss of thalamic intralaminar nuclei in progressive supranuclear palsy and Parkinson's disease: clinical and therapeutic implications. Brain 123:1410-1421

Hernán MA, Takkouche B, Caamano-Isorna F, Gestal-Otero JJ (2002) A meta-analysis of coffee drinking, cigarette smoking, and the risk of Parkinson's disease. Ann Neurol 52:276–284

Hilker R, Schweitzer K, Coburger S, Ghaerni M, Weisenbach S, Jacobs AH, Rudolf J, Herholz K, Heiss WD (2005) Nonlinear progression of Parkinson disease as determined by serial positron emission tomographic imaging of striatal fluorodopa F 18 activity. Arch Neurol 62:378–382

Hill TJ (1987) Ocular pathogenicity of herpes simplex virus. Curr Eye Res 6:1–7

Hilz MJ, Axelrod FB, Braeske K, Stemper B (2002) Cold pressor test demonstrates residual sympathetic cardiovascular activation in familial dysautonomia. J Neurol Sci 196:81–89

Hirsch EC, Graybiel AM, Duyckaerts C, Javoy-Agid F (1987) Neuronal loss in the pedunculopontine tegmental nucleus in Parkinson disease and in progressive supranuclear palsy. Proc Natl Acad Sci U S A 84:5976–5980

Hirsch EC, Graybiel AM, Agid YA (1988) Melanized dopaminergic neurons are differentially susceptible to degeneration in Parkinson's disease. Nature 334:345–347

Hirtz D, Thurman DJ, Gwinn-Hardy K, Mohamed M, Chaudhuri AR, Zalutsky R (2007) How common are the "common" neurologic disorders? Neurology 68:326–337

Hishikawa N, Hashizume Y, Yoshida M, Sobue G (2001) Widespread occurrence of argyrophilic glial inclusions in Parkinson's disease. Neuropathol Appl Neurobiol 27:362–372

Hishikawa N, Hashizume Y, Yoshida M, Niwa JI, Tanaka F, Sobue G (2005) Tuft-shaped astrocytes in Lewy body disease. Acta Neuropathol 109:373–380

Hofman A, de Jong TVM, van Duijn CM, Breteler MMB (2006) Epidemiology of neurological diseases in elderly people: what did we learn from the Rotterdam study? Lancet Neurol 5:545–550

Holst MC, Kelly JB, Powley TL (1997) Vagal preganglionic projections to the enteric nervous system characterized with Phaseolus vulgaris-leucoagglutinin. J Comp Neurol 381:81–100

Holstege G (1992) The emotional motor system. Eur J Morphol 30:67–79

Holstege G (1996) The somatic motor system. Prog Brain Res 107:9–26

Holstege G, Mouton LJ, Gerrits NM (2004) Emotional motor system. In: Paxinos G, Mai JK (eds) The human nervous system, 2nd edn. Elsevier, London, pp 1306–1325

Holt DJ, Hersh LB, Saper CB (1996) Cholinergic innervation in the human striatum: a three-compartment model. Neuroscience 74:67–87

Hopkins DA, Bieger D, de Vente J, Steinbusch HWM (1996) Vagal efferent projections: viscerotopy, neurochemistry and effects of vagotomy. Prog Brain Res 107:79–96

Hornby PJ, Abrahams TP (2000) Central control of lower esophageal sphincter relaxation. Am J Med 108:90–98

Huang XF, Paxinos G (1995) Human intermediate reticular zone: a cyto- and chemoarchitectonic study. J Comp Neurol 360:571–588

Huang XF, Törk I, Paxinos G (1993) Dorsal motor nucleus of the vagus nerve: a cyto- and chemoarchitectonic study in the human. J Comp Neurol 330:158–182

Hubbard PS, Esiri MM, Reading M, McShane R, Nagy Z (2007) Alpha-Synuclein pathology in the olfactory pathways of dementia patients. J Anat 211:117–124

Hughes AJ, Daniel SE, Kilford L, Lees AJ (1992) Accuracy of clinical diagnosis of idiopathic Parkinson's disease: a clinico-pathological study of 100 cases. J Neurol Neurosurg Psychiatry 55:181–184

Hughes AJ, Daniel SE, Ben-Shlomo Y, Lees AJ (2002) The accuracy of diagnosis of parkinsonian syndromes in a specialist movement disorder service. Brain 125:861–870

Hurtig HI, Trojanowski JQ, Galvin J, Ewbank D, Schmidt ML, Lee VMY, Clark CM, Glosser G, Stern MB, Gollomp SM, Arnold SE (2000) Alpha-synuclein cortical Lewy bodies correlate with dementia in Parkinson's disease. Neurology 54:1916–1921

Hyman BT, van Hoesen GW, Damasio AR (1990) Memory-related neural systems in Alzheimer's disease: an anatomic study. Neurology 40:1721–1730

Idiaquez J, Benarroch EE, Rosales H, Milla P, Rios L (2007) Autonomic and cognitive dysfunction in Parkinson's disease. Clin Auton Res 17:93–98

Ince PG, McKeith IG (2003) Dementia with Lewy bodies. In: Dickson DW (ed) Neurodegeneration: the molecular pathology of dementia and movement disorders. ISN Neuropathologica Press, Basel, pp 188–199

Inglis WL, Winn P (1995) The pedunculopontine tegmental nucleus: where the striatum meets the reticular formation. Prog Neurobiol 47:1–29

Insausti R, Amaral DG (2004) Hippocampal formation. In: Paxinos G, Mai KJ (eds) The human nervous system, 2nd edn. Elsevier, London, pp 872–915

Irizarry MC, Growdon W, Gómez-Isla T, Newell K, George JM, Clayton DF, Hyman BT (1998) Nigral and cortical Lewy bodies and dystrophic nigral neurites in Parkinson's disease and cortical Lewy body disease contain α-synuclein immunoreactivity. J Neuropathol Exp Neurol 57:334–337

Iseki E, Odawara T, Suzuki K, Kosaka K, Akiyama H, Ikeda K (1995) A pathological study of Lewy bodies and senile changes in the amygdala in diffuse Lewy body disease. Neuropathology 15:112–116

Issidorides MR, Mytilineou C, Panayotacopoulou T, Yahr MD (1991) Lewy bodies in parkinsonism share components with intraneuronal protein bodies of normal brains. J Neural Transm 3:49–61

Iwanaga K, Wakabayashi K, Yoshimoto M, Tomita I, Satoh H, Takashima H, Satoh A, Seto M, Tsujihata M, Takahashi H (1999) Lewy body-type degeneration in cardiac plexus in Parkinson's and incidental Lewy body diseases. Neurology 52:1269–1271

Iwatsubo T, Yamaguchi H, Fujimuro M, Yokosawa H, Ihara Y, Trojanowski JQ, Lee VM (1996) Purification and characterization of Lewy bodies from the brains of patients with diffuse Lewy body disease. Am J Pathol 148:1517–1529

Jänig W (1996) Spinal cord reflex organization of sympathetic systems. Prog Brain Res 107:43–77

Jellinger K (1988) The pedunculopontine nucleus in Parkinson's disease, progressive supranuclear palsy and Alzheimer's disease. J Neurol Neurosurg Psychiatry 51:540–543

Jellinger KA (1991) Pathology of Parkinson's disease. Changes other than the nigrostriatal pathway. Mol Chem Neuropathol 14:153–197

Jellinger KA (2001) The pathology of Parkinson's disease. Adv Neurol 86:55–72

Jellinger KA (2004) Lewy-body-related α-synucleinopathy in the aged human brain. J Neural Transm 111:1219–1235

Jellinger KA (2008) A critical reappraisal of current staging of Lewy-related pathology in human brain. Acta Neuropathol 116:1–16

Jellinger KA, Mizuno Y (2003) Parkinson's disease. In: Dickson DW (ed) Neurodegeneration: the molecular pathology of dementia and movement disorders. ISN Neuropathologica Press, Basel, pp 159–187

Jenner P (1993) Presymptomatic detection of Parkinson's disease. J Neural Transm 40:23–36

Jenner P (2003) Oxidative stress in Parkinson's disease. Ann Neurol 53:26–38

Jenner P, Olanow CW (1996) Oxidative stress and the pathogenesis of Parkinson's disease. Neurology 47:161–170

Jensen PH, Gai WP (2001) Alpha-synuclein. Axonal transport, ligand interaction, and neurodegeneration. In: Tolnay M, Probst A (eds) Neuropathology and genetics of dementia. Kluwer Academic/Plenum, New York, pp 129–134

Jones EG (1984) Laminar distribution of cortical efferent cells. In: Peters A, Jones EG (eds) Cerebral cortex, vol. 1: cellular components of the cerebral cortex. Plenum, London, pp 521–553

Jost WH, Eckardt VF (2003) Constipation in idiopathic Parkinson's disease. Scand J Gastroenterol 38:681–686

Kalaitzakis ME, Graeber MB, Gentleman SM, Pearce RKB (2007) The dorsal motor nucleus of the vagus is not an obligatory trigger site of Parkinson's disease: a critical analysis of α-synuclein staging. Neuropathol Appl Neurobiol Dec 5 (Epub ahead of print)

Kamel F, Tanner C, Umbach D, Hoppin J, Alavanja M, Blair A, Comyns K, Goldman S, Korell M, Langston J, Ross G, Sandler D (2007) Pesticide exposure and self-reported Parkinson's disease in the agricultural health study. Am J Epidemiol 165:364–374

Kapfhammer JP, Schwab ME (1994) Inverse patterns of myelination and GAP-43 expression in the adult CNS: neurite growth inhibitors as regulators of neuronal plasticity. J Comp Neurol 340:194–206

Kato S, Shaw P, Wood-Allum C, Leigh PN, Shaw C (2003) Amyotrophic lateral sclerosis. In: Dickson DW (ed) Neurodegeneration: the molecular pathology of dementia and movement disorders. ISN Neuropathologica Press, Basel, pp 350–368

Kaufmann H, Hague K, Perl D (2001) Accumulation of alpha synuclein in autonomic nerves in pure autonomic failure. Neurology 56:980–981

Kaufmann H, Nahm K, Purohit D, Wolfe D (2004) Autonomic failure as the initial manifestation of Parkinson's disease and dementia with Lewy bodies. Neurology 63:1093–1095

Kawaguchi Y, Wilson CJ, Augood SJ, Emson PC (1995) Striatal interneurones: chemical, physiological and morphological characterization. Trends Neurosci 18:527–535

Klegeris A, Pelech S, Giasson BI, Maguire J, Zhang H, McGeer E, McGeer P (2008) α-Synuclein activates stress signaling protein kinases in THP-1 cells and microglia. Neurobiol Aging 29:739-752

Klockgether T (2004) Parkinson's disease: clinical aspects. Cell Tissue Res 318:115–120

Klos KJ, Ahlskog JE, Josephs KA, Apaydin H, Parisi JE, Boeve BF, DeLucia MW, Dickson DW (2006) α-Synuclein pathology in the spinal cord of neurologically asymptomatic aged individuals. Neurology 66:1100–1102

Kövari E, Burkhardt K, Lobrinus JA, Bouras C (2007) Lewy body dysphagia. Acta Neuropathol 114:295–298

Koller WC, Montgomery EB (1997) Issues in the early diagnosis of Parkinson's disease. Neurology 49:10–25

Koller WC, Tse W (2004) Unmet medical needs in Parkinson's disease. Neurology 62:1–8

Koller WC, Langston JW, Hubble JP, Irwin I, Zack M, Golbe L, Forno L, Ellenberg J, Kurland L, Ruttenber AJ (1991) Does a long preclinical period occur in Parkinson's disease? Neurology 41:8–13

Konturek JW, Konturek SJ, Stachura J, Domschke W (1998) Helicobacter pylori-positive peptic ulcer patients do not adapt to aspirin. Aliment Pharmacol Ther 12:57–64

Kopell BH, Rezai AR, Chang JW, Vitek JL (2006) Anatomy and physiology of the basal ganglia: implications for deep brain stimulation for Parkinson's disease. Mov Disord 21:238–246

Kopito RR (2000) Aggresomes, inclusion bodies and protein aggregation. Trends Cell Biol 10:524–530

Korczyn AD (1990) Autonomic nervous system disturbances in Parkinson's disease. Adv Neurol 53:463–468

Korczyn AD (2001) Dementia in Parkinson's disease. J Neurol 248:1–4

Korczyn AD, Reichmann H (2006) Dementia with Lewy bodies. J Neurol Sci 248:3–8

Kordower JH, Chu Y, Hauser RA, Freeman TB, Olanow CW (2008) Parkinson's disease pathology in long-term embryonic nigral transplants in Parkinson's disease. Nat Med 14:504–506

Kosaka K, Yoshimura M, Ikeda K, Budka H (1984) Diffuse type of Lewy body disease: progressive dementia with abundant cortical Lewy bodies and senile changes of varying degree–a new disease? Clin Neuropathol 3:185–192

Kosaka K, Iseki E (1996) Dementia with Lewy bodies. Curr Opin Neurol 9:271–275

Koutcherov Y, Huang XF, Halliday G, Paxinos G (2004) Organisation of the human brain stem nuclei. In: Paxinos G, Mai KJ (eds) The human nervous system, 2nd edn. Elsevier, London, pp 273–321

Kramer ML, Schulz-Schaeffer WJ (2007) Presynaptic α-synuclein aggregates, not Lewy bodies, cause neurodegeneration in dementia with Lewy bodies. J Neurosci 27:1405–1410

Kranick SM, Duda JE (2008) Olfactory dysfunction in Parkinson's disease. Neurosignals 16:35–40

Kremer HP, Bots GT (1993) Lewy bodies in the lateral hypothalamus: do they imply neuronal loss? Mov Disord 8:315–320

Kubis N, Faucheux BA, Ransmayr G, Damier P, Duyckaerts C, Henin D, Forette B, Le Charpentier Y, Hauw JJ, Agid Y, Hirsch EC (2000) Preservation of midbrain catecholaminergic neurons in very old human subjects. Brain 123:366–373

Kuusisto E, Parkkinen L, Alafuzoff I (2003) Morphogenesis of Lewy bodies: dissimilar incorporation of α-synuclein, ubiquitin, and p62. J Neuropathol Exp Neurol 62:1241–1253

Lach B, Grimes D, Benoit B, Minkiewicz-Janda A (1992) Caudate nucleus pathology in Parkinson's disease: ultrastructural and biochemical findings in biopsy material. Acta Neuropathol 83:352–360

Lane RD, Reiman EM, Ahern GL, Schwartz GE, Davidson RJ (1997) Neuroanatomical correlates of happiness, sadness, and disgust. Am J Psychiatry 154:926–933

Lang AE (2007) The progression of Parkinson's disease: a hypothesis. Neurology 68:948–952

Lang AE, Lozano AM (1998a) Parkinson's disease. First of two parts. N Engl J Med 339:1044–1053

Lang AE, Lozano AM (1998b) Parkinson's disease. Second of two parts. N Engl J Med 339:1130–1143

Lang AE, Obeso JA (2004) Challenges in Parkinson's disease: restoration of the nigrostriatal dopamine system is not enough. Lancet Neurol 3:309–316

Langston JW (2006) The Parkinson's complex: parkinsonism is just the tip of the iceberg. Ann Neurol 59:591–596

Langston JW, Forno LS (1978) The hypothalamus in Parkinson's disease. Ann Neurol 3:129–133

Lantos PL, Quinn N (2003) Multiple system atrophy. In: Dickson DW (ed) Neurodegeneration: the molecular pathology of dementia and movement disorders. ISN Neuropathologica Press, Basel, pp 203–214

Lavedan C (1998) The synuclein family. Genet Res 8:871–880

Lee HG, Zhu X, Takeda A, Perry G, Smith MA (2006) Emerging evidence for the neuroprotective role of α-synuclein. Exp Neurol 200:1–7

Lee HJ, Patel S, Lee SJ (2005) Intravesicular localization and exocytosis of α-synuclein and its aggregates. J Neurosci 25:6016–6024

Lee MS, Rinne JQ, Marsden CD (2000) The pedunculopontine nucleus: its role in the genesis of movement disorders. Yonsei Med J 41:167–184

Lee VMY, Giasson BI, Trojanowski JQ (2004) More than just two peas in a pod: common amyloidogenic properties of tau and α-synuclein in neurodegenerative diseases. Trends Neurosci 27:129–134

Leigh PN, Probst A, Dale GE, Power DP, Brion JP, Dodson A, Anderton BH (1989) New aspects of the pathology of neurodegenerative disorders as revealed by ubiquitin antibodies. Acta Neuropathol 79:61–72

Lennox G, Lowe J, Landon M, Byrne EJ, Mayer RJ, Godwin-Austen RB (1989) Diffuse Lewy body disease: correlative neuropathology using anti-ubiquitin immunocytochemistry. J Neurol Neurosurg Psychiatry 52:1236–1247

Lerner A, Bagic A (2008) Olfactory Pathogenesis of idiopathic Parkinson disease revisited. Mov Disord 23:1076–1084

Leverenz JB, Umar I, Wang Q, Montine TJ, McMillan J, Tsuang DW, Jin J, Pan C, Shin J, Zhu D, Zhang J (2007) Proteomic identification of novel proteins in cortical Lewy bodies. Brain Pathol 17:139–145

Levy R, Hazrati LN, Herrero MT, Vila M, Hassani OK, Mouroux M, Ruberg M, Asensi H, Agid Y, Féger J, Obeso JA, Parent A, Hirsch EC (1997) Re-evaluation of the functional anatomy of the basal ganglia in normal and Parkinsonian states. Neuroscience 76:335–343

Lewy FH (1912) Paralysis agitans. I. Pathologische Anatomie. In: Lewandowski M (ed) Handbuch der Neurologie, vol III. Springer, Berlin Heidelberg New York, pp 920–933

Li JY, Englund E, Holton JL, Soulet D, Hagell P, Lees AJ, Lashley T, Quinn NP, Rehncrona S, Björklund A, Widner H, Revesz T, Lindvall O, Brundin P (2008) Lewy bodies in long-surviving mesencephalic grafts in Parkinson patients suggest host to graft disease propagation. Nat Med 14:501–503

Li ST, Dendi R, Holmes C, Goldstein DS (2002) Progressive loss of cardiac sympathetic innervation in Parkinson's disease. Ann Neurol 52:220–223

Liberini P, Parola S, Spano PF, Antonini I (2000) Olfaction in Parkinson's disease: methods of assessment and clinical relevance. J Neurol 247:88–96

Lippa CF (2003) Lewy bodies in conditions other than disorders of α-synuclein. In: Dickson DW (ed) Neurodegeneration: the molecular pathology of dementia and movement disorders. ISN Neuropathologica Press, Basel, pp 200–202

Lippa CF, Schmidt ML, Lee VMY, Trojanowski JQ (1999) Antibodies to α-synuclein detect Lewy bodies in many Down's syndrome brains with Alzheimer's disease. Ann Neurol 45:353–357

Litvan I, Bhatia KP, Burn DJ, Goetz CG, Lang AE, McKeith I, Quinn N, Sethi KP, Shults C, Wenning GK (2003) SIC Task force appraisal of clinical diagnostic criteria for Parkinsonian disorders. Mov Disord 18:467–486

Litvan I, Halliday G, Hallett M, Goetz CG, Rocca W, Duyckaerts C, Ben-Shlomo Y, Dickson DW, Lang AE, Chesselet MF, Langston WJ, Di Monte D, Gasser T, Hagg T, Hardy J, Jenner P, Melamed E, Myers RH, Parker D, Price D (2007) The etiopathogenesis of Parkinson's disease and suggestions for further research. J Neuropathol Exp Neurol 66:251–257

Liubashina O, Jolkkonen E, Pitkänen A (2000) Projections from the central nucleus of the amygdala to the gastric related area of the dorsal vagal complex: a Phaseolus vulgaris leucoagglutinin study in rat. Neurosci Lett 291:85–88

Loewy AD (1990) Central autonomic pathways. In: Loewy AD, Spyer KM (eds) Central regulation of autonomic functions. Oxford University Press, New York, pp 88–103

Loewy AD (1991) Forebrain nuclei involved in autonomic control. Prog Brain Res 87:253–268

Lowe J (1994) Lewy bodies. In: Calne DP (ed) Neurodegenerative diseases. Saunders, Philadelphia, pp 51–69

Loza A, Pepin JL, Rapisarda G, Moglia A, Delwaide PJ (1997) Functional changes of brainstem reflexes in Parkinson's disease. Conditioning of blink reflex R2 component by paired and index finger stimulation. J Neural Transm 104:679–687

Ludin SM, Steiger UH, Ludin HP (1987) Autonomic disturbances and cardiovascular reflexes in idiopathic Parkinson's disease. J Neurol 235:10–15

Ma SY, Roytt M, Collan Y, Rinne JQ (1999) Unbiased morphometrical measurements show loss of pigmented nigral neurones with ageing. Neuropathol Appl Neurobiol 25:394–399

Magerkurth C, Schnitzer R, Braune S (2005) Symptoms of autonomic failure in Parkinson's disease: prevalence and impact on daily life. Clin Auton Res 15:76–82

Marsden CD, Obeso JA (1994) The functions of the basal ganglia and the paradox of stereotaxic surgery in Parkinson's disease. Brain 117:877–897

Martignoni E, Pacchetti C, Godi L, Miceli G, Nappi G (1995) Autonomic disorders in Parkinson's disease. J Neural Transm 45:11–19

Martignoni E, Tassorelli C, Nappi G (2006) Cardiovascular dysautonomia as a cause of falls in Parkinson's disease. Parkinsonism Relat Disord 12:195–204

Mathias CJ (2002) To stand on one's own legs. Clin Med 2:237–245

Matsuda K, Park CH, Synden Y, Kimura T, Ochiai K, Kida H, Umemura T (2004) The vagus nerve is one route of transneural invasion for intranasally inoculated influenza A virus in mice. Vet Pathol 41:101–107

Matsumura M, Kojima J (2001) The role of the pedunculopontine tegmental nucleus in experimental parkinsonism in primates. Stereotact Funct Neurosurg 77:108–115

Mattila PM, Rinne JO, Helenius H, Dickson DW, Röyttä M (2000) Alpha-synuclein-immunoreactive cortical Lewy bodies are associated with cognitive impairment in Parkinson's disease. Acta Neuropathol 100:285–290

Matzuk MM, Saper CB (1985) Preservation of hypothalamic dopaminergic neurons in Parkinson's disease. Ann Neurol 18:552–555

McAlonan K, Brown VJ (2002) The thalamic reticular nucleus: more than a sensory nucleus? Neuroscientist 8:302–305

McBride PA, Schulz-Schaeffer WJ, Donaldson M, Bruce M, Diringer H, Kretzschmar HA, Beekes M (2001) Early spread of scrapie from the gastrointestinal tract to the central nervous system involves autonomic fibers of the splanchnic and vagus nerves. J Virol 75:9320–9327

McDonald WM, Richard IH, DeLong MR (2003) Prevalence, etiology, and treatment of depression in Parkinson's disease. Biol Psychiatry 54:363–375

McGeer PL, McGeer EG (2008) The α-synuclein burden hypothesis of Parkinson disease and its relationship to Alzheimer disease. Exp Neurol 212:235–238

McKeith IG, Dickson DW, Lowe J, Emre E, O'Brien JT, Feldman H, Cummings J, Duda JE, Lippa C, Perry EK, Aarsland D, Arai H, Ballard CG, Boeve B, Burn DJ, Costa D, Del Ser T, Dubois B, Galasko D, Gauthier S, Goetz CG, Gómez-Tortosa E, Halliday G, Hansen LA, Hardy J, Iwatsubo T, Kalaria RN, Kaufer D, Korczyn A, Kosaka K, Lee VM-Y, Lees A, Litvan I, Londos E, Lopez OL, Minoshima S, Mizuno Y, Molina JA, Mukaetova-Ladinska EB, Pasquier F, Perry RH, Schulz JB, Trojanowski JQ, Yamada M (2005) Diagnosis and management of dementia with Lewy bodies: third report of the DLB consortium. Neurology 65:1863–1872

McNaught KSP, Jenner P (2001) Proteasomal function is impaired in substantia nigra in Parkinson's disease. Neurosci Lett 297:191–194

McNaught KSP, Shashidharan P, Perl DP, Jenner P, Olanow CW (2002) Aggresome-related biogenesis of Lewy bodies. Eur J Neurosci 16:2136–2148

McNeill TH, Brown SA, Rafols JA, Shoulson I (1988) Atrophy of medium spiny I striatal dendrites in advanced Parkinson's disease. Brain Res 455:148–152

McRitchie DA, Cartwright HR, Halliday GM (1997) Specific A10 dopaminergic nuclei in the midbrain degenerate in Parkinson's disease. Exp Neurol 144:202–213

Meco G, Pratesi L, Bonifati V (1991) Cardiovascular reflexes and autonomic dysfunction in Parkinson's disease. J Neurol 238:195–199

Mena-Segovia J, Bolam JP, Magill PJ (2004) Pedunculopontine nucleus and basal ganglia: distant relatives or part of the same family? Trends Neurosci 27:585–588

Meredith G, Sonsalla P, Chesselet MF (2008) Animal models of Parkinson's disease progression. Acta Neuropathol (in press)

Meredith GE, Pattiselanno A, Groenewegen HJ, Haber SN (1996) Shell and core in monkey and human nucleus accumbens identified with antibodies to calbindin-D28k. J Comp Neurol 365:628–639

Mesholam RL, Moberg PJ, Mahr RN, Doty RL (1998) Olfaction in neurodegenerative disease. A meta-analysis of olfactory functioning in Alzheimer's and Parkinson's diseases. Arch Neurol 55:84–90

Mesulam MM (1998) From sensation to cognition. Brain 121:1013–1052

Mesulam MM (2004) The cholinergic innervation of the human cerebral cortex. Prog Brain Res 145:67–78

Mesulam MM, Mufson EJ (1993) The insula of Reil in man and monkey. In: Jones EG, Peters A (eds) Cerebral cortex. Association and auditory cortices, vol 4. Plenum, London, pp 179–225

Mesulam MM, Geula C, Bothwell MA, Hersh LB (1989) Human reticular formation: cholinergic neurons of the pedunculopontine and laterodorsal tegmental nuclei and some cytochemical comparisons to forebrain cholinergic neurons. J Comp Neurol 283:611–633

Mesulam MM, Hersh LB, Mash DC, Geula C (1992) Differential cholinergic innervation within functional subdivisions of the human cerebral cortex—a choline acetyltransferase study. J Comp Neurol 318:316–328

Mikolaenko I, Pletnikova O, Kawas CH, O'Brien R, Resnick SM, Crain B, Troncoso JC (2005) Alpha-synuclein lesions in normal aging, Parkinson disease, and Alzheimer disease: evidence from the Baltimore Longitudinal Study of Aging (BLSA). J Neuropathol Exp Neurol 64:156–162

Miller JA, Oldham MC, Geschwind DH (2008) A system level analysis of trascriptional changes in Alzheimer's disease and normal aging. J Neurosci 28:1410-1420

Minguez-Castellanos A, Chamorro CE, Escamilla-Sevilla F, Ortega-Moreno A, Rebollo AC, Gomez-Rio M, Concha A, Munoz DG (2007) Do α-synuclein aggregates in autonomic plexuses predate Lewy body disorders? A cohort study. Neurology 68:2012–2018

Miwa H, Kubo T, Suzuki A, Kondo T (2006) Intragastric proteasome inhibition induces alpha-synuclein-immunopositive aggregations in neurons in the dorsal motor nucleus of the vagus in rats. Neurosci Lett 401:146–149

Moceri VM, Kukull WA, Emanuel I, van Belle G, Larson EB (2000) Early-life risk factors and the development of Alzheimer's disease. Neurology 54:415–420

Mollenhauer B, Cullen V, Kahn I, Krastins B, Outeiro TF, Pepivani I, Ng J, Schulz-Schaeffer W, Kretzschmar HA, McLean PJ, Trenkwalder C, Sarracino DA, Vonsattel JP, Locascio JJ, El-Agnaf OM, Schlossmacher MG (2008) Direct quantification of CSF alpha-synuclein by ELISA and first cross-sectional study in patients with neurodegeneration. Exp Neurol 213:315–325

Moran LB, Hickey L, Michael GJ, Derkacs M, Christian LM, Kalaitzakis ME, Pearce RKB, Graeber M (2007) Neuronal pentraxin II is highly upregulated in Parkinson's disease and a novel component of Lewy bodies. Acta Neuropathol 115:471–478

Morecraft RJ, Geula C, Mesulam MM (1993) Architecture of connectivity within a cingulo-fronto-parietal neurocognitive network for directed attention. Arch Neurol 50:279–284

Morens DM, Davis JW, Grandinetti A, Ross GW, Popper JS, White LR (1996) Epidemiologic observations on Parkinson's disease: incidence and mortality in a prospective study of middle-aged men. Neurology 46:1044–1050

Mori F, Tanji K, Yoshimoto M, Takahashi H, Wakabayashi K (2002) Demonstration of α-synuclein immunoreactivity in neuronal and glial cytoplasm in normal human brain tissue using proteinase K and formic acid pretreatment. Exp Neurol 176:98–104

Mori F, Tanji K, Zhang H, Kakita A, Takahashi H, Wakabayashi K (2007) α-Synuclein pathology in the neostriatum in Parkinson's disease. Acta Neuropathol [Epub 2007 Nov 14]

Morrish PK, Sawle GV, Brooks DJ (1996) An [18F]dopa-PET and clinical study of the rate of progression in Parkinson's disease. Brain 119:585–591

Morrish PK, Rakshi JS, Bailey DL, Sawle GV, Brooks DJ (1998) Measuring the rate of progression and estimating the preclinical period of Parkinson's disease with [18F]dopa PET. J Neurol Neurosurg Psychiatry 64:314–319

Morrison LA, Sidman RL, Fields BN (1991) Direct spread of reovirus from the intestinal lumen to the central nervous system through vagal autonomic nerve fibers. Proc Natl Acad Sci USA 88:3852–3856

Müller CM, de Vos RAI, Maurage CA, Thal DR, Tolnay M, Braak H (2005) Staging of sporadic Parkinson disease-related alpha-synuclein pathology: inter- and intra-rater reliability. J Neuropathol Exp Neurol 64:623–628

Müller J, Wenning GK, Verny M, McKee A, Chaudhuri KR, Jellinger K, Poewe W, Litvan I (2001) Progression of dysarthria and dysphagia in postmortem-confirmed parkinsonian disorders. Arch Neurol 58:259–264

Münch G, Lüth HJ, Wong A, Arendt T, Hirsch E, Ravid R, Riederer P (2000) Crosslinking of α-synuclein by advanced glycation endproducts—an early pathophysiological step in Lewy body formation? J Chem Neuroanat 20:253–257

Nakano K (2000) Neural circuits and topographic organization of the basal ganglia and related regions. Brain Dev 22:5–16

Neafsy EJ (1990) Prefrontal cortical control of the autonomic nervous system: anatomical and physiological observations. Prog Brain Res 85:147–166

Neely MD, Schmidt DE, Deutch AY (2007) Cortical regulation of dopamine depletion-induced dendritic spine loss in striatal medium spiny neurons. Neuroscience 149:457–464

Nelson DA, Paulson GW (2002) Idiopathic Parkinson's disease(s) may follow subclinical episodes of perivenous demyelination. Med Hypotheses 59:762–769

Neumann M, Müller V, Kretzschmar HA, Haass C, Kahle PJ (2004) Regional distribution of proteinase-K-resistant α-synuclein correlates with Lewy body disease stage. J Neuropathol Exp Neurol 63:1225–1235

Nicotera P (2001) A route for prion neuroinvasion. Neuron 31:345–348

Nieuwenhuys R (1996) The greater limbic system, the emotional motor system and the brain. Prog Brain Res 107:551–580

Nieuwenhuys R (1999) Structure and organisation of fibre systems. In: Nieuwenhuys R, Ten Donkelaar HJ, Nicholson C (eds) The central nervous system of vertebrates, vol 1. Springer, Berlin Heidelberg New York, pp 113–157

Norris EH, Giasson BI, Lee VM (2004) α-Synuclein: normal function and role in neurodegenerative diseases. Curr Top Dev Biol 60:17–54

Nussbaum RL, Polymeropoulos MH (1997) Genetics of Parkinson's disease. Hum Mol Genet 6:1687–1691

Obeso JA, Guridi J, Obeso JA, DeLong M (1997) Surgery for Parkinson's disease. J Neurol Neurosurg Psychiatry 62:2–8

Öngür D, An X, Price JL (1998) Prefrontal cortical projections to the hypothalamus in macaque monkeys. J Comp Neurol 401:480–505

Ohm TG, Heilmann R, Braak H (1989) The human oral raphe system. Architectonics and neuronal types in pigment-Nissl-preparations. Anat Embryol 180:37–43

Olanow CW (1992) An introduction to the free radical hypothesis in Parkinson's disease. Ann Neurol 32:2–9

Olanow CW, McNaught KS (2006) Ubiquitin-proteasome system and Parkinson's disease. Mov Disord 21:1806–1823

Olanow CW, Perl DP, DeMartino GN, McNaught KSP (2004) Lewy-body formation is an aggresome-related process: a hypothesis. Lancet Neurol 3:496–503

Oppenheimer SM, Gelb A, Girvin JP, Hachinski VC (1992) Cardiovascular effects of human insular cortex stimulation. Neurology 42:1727–1732

Orimo S, Amino T, Itoh Y, Takahashi A, Kojo T, Uchihara T, Tsuchiya K, Mori F, Wakabayashi K, Takahashi H (2005) Cardiac sympathetic denervation precedes neuronal loss in the sympathetic ganglia in Lewy body disease. Acta Neuropathol 109:583–588

Orimo S, Takahashi A, Uchihara T, Mori F, Kakita A, Wakabayashi K, Takahashi H (2007) Degeneration of cardiac sympathetic nerve begins in the early disease process of Parkinson's disease. Brain Pathol 17:24–30

Pahapill PA, Lozano AM (2000) The pedunculopontine nucleus and Parkinson's disease. Brain 123:1767–1783

Palka-Santini M, Schwarz-Henke B, Hösel M, Renz D, Auerochs S, Brondke H, Doerfler W (2003) The gastrointestinal tract as the portal of entry for foreign macromolecules: fate of DANN and proteins. Mol Genet Genomics 270:201–215

Pandya DN, Yeterian EH (1985) Architecture and connections of cortical association areas. In: Jones EG, Peters A (eds) Cerebral cortex. Association and auditory cortices, vol 4. Plenum, London, pp 3–61

Pandya DN, Yeterian EH (1993) Architecture and connections of cerebral cortex: implications for brain evolution and function. In: Scheibel AB, Wechsler AF (eds) Neurobiology of higher function. Guilford Press, New York, pp 53–84

Pandya DN, Yeterian EH (1996) Comparison of prefrontal architecture and connections. Philos Trans R Soc Lond B Biol Sci 351:1423-1432

Panula P, Airaksinen MS, Pirvola U, Kotilainen E (1990) A histamine-containing neuronal system in human brain. Neuroscience 34:127–132

Parent A, Cicchetti F (1998) The current model of basal ganglia organization under scrutiny. Mov Disord 13:199–202

Parent A, Hazrati LN (1995a) Functional anatomy of the basal ganglia. I. The cortico-basal ganglia-thalamo-cortical loop. Brain Res Brain Res Rev 20:91–127

Parent A, Hazrati LN (1995b) Functional anatomy of the basal ganglia. II. The place of subthalamic nucleus and external pallidum in basal ganglia circuitry. Brain Res Brain Res Rev 20:128–154

Parent A, Sato F, Wu Y, Gauthier J, Lévesque M, Parent M (2000) Organization of the basal ganglia: the importance of axonal collateralization. Trends Neurosci 23:20–27

Parkinson JD (1817) An essay on the shaking palsy. Sherwood Neely and Jones, London

Parkkinen L, Soininen H, Alafuzoff I (2003) Regional distribution of alpha-synuclein pathology in unimpaired aging and Alzheimer's disease. J Neuropathol Exp Neurol 62:363–367

Parkkinen L, Kauppinen T, Pirttilä T, Autere JM, Alafuzoff I (2005) Alpha-synuclein pathology does not predict extrapyramidal symptoms of dementia. Ann Neurol 57:82–91

Parkkinen L, Pirttilä T, Alafuzoff I (2008) Applicability of current staging systems of α-synuclein pathology and their clinical relevance. Acta Neuropathol 115:399–407

Pasik P, Pasik T, DiFiglia M (1976) Quantitative aspects of neuronal organization in the neostriatum of the macaque monkey. In: Yahr MD (ed) The basal ganglia. Raven Press, New York, pp 57–90

Passani MB, Bacciottini L, Mannaioni F, Blandina P (2000) Central histaminergic system and cognition. Neurosci Biobehav Rev 24:107–113

Pearce RK, Hawkes CH, Daniel SE (1995) The anterior olfactory nucleus in Parkinson's disease. Mov Disord 10:283–287

Pearson RCA (1996) Cortical connections and the pathology of Alzheimer's disease. Neurodegeneration 5:429–434

Pearson RCA, Esiri MM, Hiorns RW, Wilcock GK, Powell TPS (1985) Anatomical correlates of the distribution of the pathological changes in the neocortex in Alzheimer's disease. Proc Natl Acad Sci USA 82:4531–4534

Perrin RJ, Woods WS, Clayton DF, George JM (2000) Interaction of human alpha-synuclein and Parkinson's disease variants with phospholipids. J Biol Chem 44:34393–34398

Perry RH, Jaros EB, Irving D, Scones DJ, Brown A, McMeekin WM (1996) What is the neuropathological basis of dementia associated with Lewy bodies? In: Perry RH, McKeith IG, Perry EK (eds) Dementia with Lewy bodies. Cambridge University Press, New York, pp 212–223

Petrides M, Pandya DN (2004) The frontal cortex. In: Paxinos G, Mai JK (eds) The human nervous system, 2nd edn. Elsevier, London, pp 951–974

Petrovitch H, Ross GW, Abbott RD, Sanderson WT, Sharp DS, Tanner CM, Masaki KH, Blanchette PL, Popper JS, Foley D, Launer L, White LR (2002) Plantation work and risk of Parkinson disease in a population-based longitudinal study. Arch Neurol 59:1787–1792

Peuralinna T, Oinas M, Polvikoski T, Paetau A, Sulkava R, Niinistö L, Kalimo H, Hernandez D, Hardy J, Singleton A, Tienari PJ, Myllykangas L (2008) Neurofibrillary tau pathology modulated by genetic variation of α-synuclein Ann Neurol 64:348–352

Pfeiffer RF (1998) Gastrointestinal dysfunction in Parkinson's disease. Clin Neurosci 5:136–146

Pfeiffer RF (2003) Gastrointestinal dysfunction in Parkinson's disease. Lancet Neurol 2:107–116

Phillips RJ, Powley TL (2007) Innervation of the gastrointestinal tract: pattern of aging. Auton Neurosci 136:1–19

Phillips RJ, Walter GC, Wilder SL, Baronowski EA, Powley TL (2008) Alpha-synuclein immunopositive myenteric neurons and vagal preganglionic terminals: autonomic pathway implicated in Parkinson's disease? Neuroscience (in press)

Piao YS, Wakabayashi K, Hayashi S, Yoshimoto M, Takahashi H (2000) Aggregation of α-synuclein/NACP in the neuronal and glial cells in diffuse Lewy body disease: a survey of six patients. Clin Neuropathol 19:163–169

Pillon B, Deweer B, Malapani C, Rogelet P, Agid Y, Dubois B (1994) Explicit memory disorders of demented parkinsonian patients and underlying neuronal basis. In: Korczyn AD (ed) Dementia in Parkinson's disease. Monduzzi Editore, Bologna, pp 265–271

Pletnikova O, West N, Lee MK, Rudow GL, Skolasky RL, Dawson TM, Marsh L, Troncoso JC (2005) Aβ deposition is associated with enhanced cortical α-synuclein lesions in Lewy body diseases. Neurobiol Aging 26:1183–1192

Pollanen MS, Dickson DW, Bergeron C (1993) Pathology and biology of the Lewy body. J Neuropathol Exp Neurol 52:183–191

Ponsen MM, Stoffers D, Booij J, van Eck-Smit BL, Wolters EC, Berendse HW (2004) Idiopathic hyposmia as a preclinical sign of Parkinson's disease. Ann Neurol 56:173–181

Porta EA (2002) Pigments in aging: and overview. Ann N Y Acad Sci 959:57–65

Price JL (2004) Olfaction. In: Paxinos G, Mai JM (eds) The human nervous system, 2nd edn. Elsevier, London, pp 1198–1212

Price JL, Russchen FT, Amaral DG (1987) The amygdaloid complex. In: Björklund A, Hökfelt T, Swansen LW (eds) Handbook of chemical neuroanatomy, vol. 5(1): integrated systems. Elsevier, Amsterdam, pp 279–388

Price JL, Carmichael ST, Drevets WC (1996) Networks related to the orbital and medial prefrontal cortex: a substrate for emotional behavior? Prog Brain Res 107:528–536

Przuntek H, Müller T, Riederer P (2004) Diagnostic staging of Parkinson's disease: conceptual aspects. J Neural Transm 111:201–216

Qualman SJ, Haupt HM, Yang P, Hamilton SR (1984) Esophageal Lewy bodies associated with ganglion cell loss in achalasia. Similarity to Parkinson's disease. Gastroenterology 87:848–856

Quigley EM (1996) Gastrointestinal dysfunction in Parkinson's disease. Semin Neurol 16:245–250

Rajput AH (1994) Clinical features and natural history of Parkinson's disease (special consideration of aging). In: Calne DP (ed) Neurodegenerative diseases. Saunders, Philadelphia, pp 555–571

Rajput AH, Rozdilsky B (1976) Dysautonomia in Parkinsonism: a clinicopathological study. J Neurol Neurosurg Psychiatry 29:1092-1100

Randich A, Gebhart GF (1992) Vagal afferent modulation of nociception. Brain Res Brain Res Rev 17:77–99

Rascol O, Payoux P, Ory F, Ferreira JJ, Brefel-Courbon C, Montastruc JL (2003) Limitations of current Parkinson's disease therapy. Ann Neurol 53:3–15

Reiman EM, Lane RD, Ahern GL, Schwartz GE, Davidson RJ, Friston KJ, Yun LS, Chen K (1997) Neuroanatomical correlates of externally and internally generated human emotion. Am J Psychiatry 154:918–925

Reiner A, Anderson KD (1990) The patterns of neurotransmitter and neuropeptide co-occurrence among striatal projection neurons: conclusions based on recent findings. Brain Res Brain Res Rev 15:251–265

Reisberg B, Pattschull-Furlan A, Franssen E, Sclan SG, Kluger A, Dingcong L, Ferris SH (1992) Dementia of the Alzheimer type recapitulates ontogeny inversely on specific ordinal and temporal parameters. In: Kostovic I, Knezevic S, Wisniewski HM, Spillich GJ (eds) Neurodevelopment, aging and cognition. Birkhäuser, Boston, pp 345–369

Reisberg B, Franssen EH, Hasan SM, Monteiro I, Boksay I, Souren LEM, Kenowsky S, Auer SR, Elahi S, Kluger A (2003) Retrogenesis: clinical, physiologic, and pathologic mechanisms in brain aging, Alzheimer's and other dementing processes. Eur Arch Psychiatry Clin Neurosci 249:28–36

Revesz T, Daniel SE (1998) Corticobasal degeneration. In: Markesbery WR (ed) Neuropathology of dementing disorders. Arnold, New York, pp 257–267

Riess O, Krüger R, Hochstrasser H, Soehn AS, Nuber S, Franck T, Berg D (2006) Genetic causes of Parkinson's disease: extending the pathway. J Neural Transm 70:181–189

Rinaman L, Levitt P, Card JP (2000) Progressive postnatal assembly of limbic-autonomic circuits revealed by central transneuronal transport of pseudorabies virus. J Neurosci 20:2731–2741

Rinne JO, Rummukainen J, Paljärvi L, Rinne UK (1989) Dementia in Parkinson's disease is related to neuronal loss in the medial substantia nigra. Ann Neurol 26:47–50

Rockland KS, Pandya DN (1979) Laminar origins and terminations of cortical connections of the occipital lobe in the rhesus monkey. Brain Res 179:3–20

Ross GW, Abbott RD, Petrovitch H, Tanner CM, Davis DG, Nelson J, Markesbery WR, Hardman J, Masaki K, Launer L, White LR (2006) Association of olfactory dysfunction with incidental Lewy bodies. Mov Disord 21:2062–2067

Ross GW, Petrovitch H, Abbott RD, Tanner CM, Popper Masaki K (2008) Association of olfactory dysfunction with risk of future Parkinson's disease. Ann Neurol 63:167–173

Rossi A, Giovenali P, Benvenuti M, Di Iorio W, Calabresi P (2007) Skin biopsy: a new diagnostic tool for autonomic dysfunctions in Parkinson's disease? Lancet Neurol 6:848–849

Royce GJ (1982) Laminar origin of cortical neurons which project upon the caudate nucleus: a horseradish peroxidase investigation in the cat. J Comp Neurol 205:8–29

Rudow G, O'Brien R, Savonenko A, Resnick SM, Zonderman AB, Pletnikova O, Marsh L, Dawson T, Crain B, West M, Troncoso JC (2008) Morphometry of the human substantia nigra in ageing and Parkinson's disease. Acta Neuropathol 115:461–470

Rüb U, Del Tredici K, Schultz C, Ghebremedhin E, de Vos RAI, Jansen Steur E, Braak H (2002) Parkinson's disease: the thalamic components of the limbic loop are severely impaired by α-synuclein immunopositive inclusion body pathology. Neurobiol Aging 23:245–254

Rye DB (1997) Contributions of the pedunculopontine region to normal and altered REM sleep. Sleep 29:757–788

Sabin TD (1996) Clinical implications of retrograde neuronal suicide transport. Neurologist 2:176–184

Sage JI (2004) Pain in Parkinson's disease. Curr Treat Options Neurol 6:191–200

Saha AR, Hill J, Utton MA, Asuni AA, Ackerley S, Grierson AJ, Miller CC, Davies AL, Buchman VL, Anderton BH, Hanger DP (2004) Parkinson's disease α-synuclein mutations exhibit defective axonal transport in cultured neurons. J Cell Sci 117:1017–1024

Saito Y, Kawashima A, Ruberu NN, Fujiwara H, Koyama S, Sawabe M, Arai T, Nagura H, Yamanouchi H, Hasegawa M, Iwatsubo T, Murayama S (2003) Accumulation of phosphorylated α-synuclein in aging human brain. J Neuropathol Exp Neurol 62:644–654

Saito Y, Ruberu NN, Sawabe M, Arai T, Kazama H, Hosoi T, Yamanouchi H, Murayama S (2004) Lewy body-related α-synucleinopathy in aging. J Neuropathol Exp Neurol 63:742–749

Sakuma K, Nakashima K, Takahashi K (1996) Olfactory evoked potentials in Parkinson's disease, Alzheimer's disease and anosmic patients. Psychiatry Clin Neurosci 50:35–40

Sandmann-Keil D, Braak H, Okochi M, Haass C, Braak E (1999) Alpha-synuclein immunoreactive Lewy bodies and Lewy neurites in Parkinson's disease are detectable by an advanced silver-staining technique. Acta Neuropathol 98:461–464

Sanides F (1969) Comparative architectonics of the neocortex of mammals and their evolutionary interpretation. Ann N Y Acad Sci 167:404–423

Saper CB (1987) Diffuse cortical projection systems: anatomical organization and role in cortical function. In: Plum F (ed) Handbook of physiology. The nervous system, vol V. Am Physiol Soc, Bethesda, pp 169–210

Saper CB (1990) Cholinergic system. In: Paxinos G (ed) The human nervous system. Academic Press, San Diego, pp 1095–1113

Saper CB (1996) Role of the cerebral cortex and striatum in emotional motor response. Prog Brain Res 107:537–550

Saper CB (1998) "All fall down": the mechanism of orthostatic hypotension in multiple systems atrophy and Parkinson's disease. Ann Neurol 43:149–151

Saper CB (2004) Hypothalamus. In: Paxinos G, Mai JK (eds) The human nervous system, 2nd edn. Elsevier, London, pp 514–550

Saper CB, Petito CK (1982) Correspondence of melanin-pigmented neurons in human brain with A1-A14 catecholamine cell groups. Brain 105:87–101

Saper CB, Wainer BH, German DC (1987) Axonal and transneuronal transport in the transmission of neurological disease: potential role in system degenerations, including Alzheimer's disease. Neuroscience 23:389–398

Saper CB, Sorrentino DM, German DC, de Lacalle S (1991) Medullary catecholaminergic neurons in the normal human brain and in Parkinson's disease. Ann Neurol 29:577–584

Sawle GV (1993) The detection of preclinical Parkinson's disease: what is the role of positron emission tomography? Mov Disord 8:271–277

Scherder E, Wolters E, Polman C, Serfeant J, Swaab D (2005) Pain in Parkinson's disease and multiple sclerosis: its relation to the medial and lateral pain systems. Neurosci Biobehav Rev 29:1047–1056

Schwizer W, Steingoetter A, Fox M (2006) Magnetic resonance imaging for the assessment of gastrointestinal functions. Scand J Gastroenterol 41:1245–1260

Selemon LD, Goldman-Rakic PS (1985) Longitudinal topography and interdigitating of corticostriatal projections in the rhesus monkey. J Neurosci 5:776–794

Sethi K (2008) Levodopa unresponsive symptoms in Parkinson disease. Mov Disord 23 (suppl 3): S521–S533

Sharabi Y, Li ST, Dendi R, Holmes C, Goldstein DS (2003) Neurotransmitter specificity of sympathetic denervation in Parkinson's disease. Neurology 60:1036–1039

Sherer TB, Betarbet R, Testa CM, Seo BB, Richardson JR, Kim JH, Miller GW, Yagi T, Matsuno-Yagi A, Greenamyre JT (2003) Mechanism of toxicity in rotenone models of Parkinson's disease. J Neurosci 23:10756–10764

Sherk H (1986) The claustrum and the cerebral cortex. In: Jones EG, Peters A (eds) Cervebral cortex. sensory-motor areas and aspects of cortical connectivity. vol 5. Plenum, London, pp 467–499

Shults CW (2006) Lewy bodies. Proc Natl Acad Sci USA 103:1661–1668

Sibon I, Tison F (2004) Vascular parkinsonism. Curr Opin Neurol 17:49–54

Siddiqui MF, Rast S, Lynn MJ, Auchus AP, Pfeiffer RF (2002) Autonomic dysfunction in Parkinson's disease: a comprehensive symptom survey. Parkinsonism Rel Disord 8:277–284

Sims KS, Williams RS (1990) The human amygdaloid complex: a cytologic and histochemical atlas using Nissl, myelin, acetylcholinesterase and nicotinamide adenine dinucleotide phosphate diaphorase staining. Neuroscience 36:449–472

Singer C, Weiner WJ, Sanchez-Ramos JR (1992) Autonomic dysfunction in men with Parkinson's disease. Eur Neurol 32:134–140

Smith AD, Bolam JP (1990) The neural network of the basal ganglia as revealed by the study of synaptic connections of identified neurones. Trends Neurosci 13:259–265

Smith Y, Shink E, Sidibé M (1998) Neuronal circuitry and synaptic connectivity of the basal ganglia. Neurosurg Clin N Am 9:203–222

Snow BJ (1996) Fluorodopa PET scanning in Parkinson's disease. Adv Neurol 69:449–457

Sommer U, Hummel T, Cormann K, Mueller A, Frasnelli J, Kropp J, Reichmann H (2004) Detection of presymptomatic Parkinson's disease: combining smell tests, transcranial sonography, and SPECT. Mov Disord 19:1196–1202

Sotelo C, Triller A (1997) The central neuron. In: Graham DI, Lantos PL (eds) Greenfield's neuropathology, 6th edn. Arnold, London, pp 3–62

Soykan I, Lin Z, Bennet JP, McCallum RW (1999) Gastric myelectrical activity in patients with Parkinson's disease: evidence of a primary gastric abnormality. Dig Dis Sci 44:927–931

Spillantini MG, Schmidt ML, Lee VMY, Trojanowski JQ, Jakes R, Goedert M (1997) α-Synuclein in Lewy bodies. Nature 388:839–840

Spillantini MG, Crowther RA, Jakes R, Cairns NJ, Lantos PL, Goedert M (1998) Filamentous alpha-synuclein inclusions link multiple system atrophy with Parkinson's disease and dementia with Lewy bodies. Neurosci Lett 31:205–208

Squire LR, Schacter DL (2002) Neuropsychology of memory. The Guilford Press, New York

Starr MS (1995) Glutamate/dopamine D1/D2 balance in the basal ganglia and its relevance to Parkinson's disease. Synapse 19:264–293

Steckler T, Inglis W, Winn P, Sahgal A (1994) The pedunculopontine tegmental nucleus: a role in cognitive processes? Brain Res Brain Res Rev 19:298–318

Stefani A, Lozano AM, Peppe A, Stanzione P, Galati S, Tropepi D, Pierantozzi M, Scarnati E, Mazzone P (2007) Bilateral deep brain stimulation of the pedunculopontine and subthalamic nuclei in severe Parkinson's disease. Brain 130:1596–1607

Stephens B, Mueller AJ, Shering AF, Hood SH, Taggart P, Arbuthnott GW, Bell JE, Kilford L, Kingsbury AE, Daniel SE, Ingham CA (2005) Evidence of a breakdown of corticostriatal connections in Parkinson's disease. Neuroscience 132:741–754

Stiasny-Kolster K, Doerr Y, Möller JC, Hoffken H, Behr TM, Oertel WH, Mayer G (2005) Combination of 'idiopathic' REM sleep behaviour disorder and olfactory dysfunction as possible indicator for α-synucleinopathy demonstrated by dopamine transporter FP-CIT-SPECT. Brain 128:126–137

Stolze H, Kuhtz-Buschbeck JP, Drücke H, Jöhnk K, Illert M, Deuschl G (2001) Comparative analysis of the gait disorder of normal pressure hydrocephalus and Parkinson's disease. J Neurol Neurosurg Psychiatry 70:289–297

Stolzing A, Grune T (2001) The proteasome and its function in the ageing process. Clin Exp Dermatol 26:566–572

Strack AM, Sawyer WB, Hughes JH, Platt KB, Loewy AD (1989) A general pattern of CNS innervation of the sympathetic outflow demonstrated by transneuronal pseudorabies viral infections. Brain Res 491:156–162

Strang RR (1965) The association of gastro-duodenal ulceration with Parkinson's disease. Med J Aust 52:842–843

Sun MK (1995) Central neural organization and control of sympathetic nervous system in mammals. Prog Neurobiol 47:157–233

Sung JY, Park SM, Lee CH, Um JW, Lee HJ, Kim J, Oh YJ, Lee ST, Paik SR, Chung KC (2005) Proteolytic cleavage of extracellular secreted α-synuclein via matrix metalloproteinases. J Biol Chem 280:25216–25224

Suzuki WA, Amaral DG (2004) Functional neuroanatomy of the medial temporal lobe memory system. Cortex 40:220–222

Takahashi H, Wakabayashi K (2001) The cellular pathology of Parkinson's disease. Neuropathology 21:315–322

Takahashi H, Wakabayashi K (2005) Controversy: is Parkinson's disease a single disease entity? Yes. Parkinsonism Rel Disord 11:31–37

Takatsu H, Nishida H, Matsuo H, Watanabe S, Nagashima K, Wada H, Noda T, Nishigaki K, Fujiwara H (2000) Cardiac sympathetic denervation from the early stage of Parkinson's disease: clinical and experimental studies with radiolabeled MIBG. J Nucl Med 41:71–77

Takeda A, Hashimoto M, Mallory M, Sundsumo M, Hansen L, Sisk A, Masliah E (1998) Abnormal distribution of the non-Aβ component of Alzheimer's disease amyloid precursor/α-synuclein in Lewy body disease as revealed by proteinase K and formic acid pretreatment. Lab Invest 78:1169–1177

Takeda A, Hashimoto M, Mallory M, Sundsumo M, Hansen L, Masliah E (2000) C-terminal α-synuclein immunoreactivity in structures other than Lewy bodies in neurodegenerative disorders. Acta Neuropathol 99:296–304

Takeda S, Yamazaki K, Miyakawa T, Arai H (1993) Parkinson's disease with involvement of the parasympathetic ganglia. Acta Neuropathol 86:397–398

Taki J, Yoshita M, Yamada M, Tonami N (2004) Significance of 123J-MIBG scintigraphy as a pathophysiological indicator in the assessment of Parkinson's disease and related disorders: it can be a specific marker for Lewy body disease. Ann Nucl Med 18:453–461

Taly AB, Meenakshi-Sundaram S, Sinha S, Swamy HS, Arunodaya GR (2007) Wilson disease: description of 282 patients evaluated over 3 decades. Medicine (Baltimore) 86:112–121

Tanner CM, Aston DA (2000) Epidemiology of Parkinson's disease and akinetic syndromes. Curr Opin Neurol 13:427–430

Tanner CM, Goetz CG, Klavans HL (1992) Autonomic nervous system disorders in Parkinson's disease. In: Koller WC (ed) Handbook of Parkinson's disease, 2nd edn. Marcel Dekker, London, pp 185–215

Tanner CM, Goldman SM, Aston DA, Ottman R, Ellenberg J, Mayeux R, Langston JW (2002) Smoking and Parkinson's disease in twins. Neurology 58:581–588

Temel Y, Visser-Vandewalle V (2006) Targets for deep brain stimulation in Parkinson's disease. Expert Opin Ther Targets 10:1–8

Tepper JM, Bolam JP (2004) Functional diversity and specificity of neostriatal interneurons. Curr Opin Neurobiol 14:685–692

Tepper JM, Abercrombie ED, Bolam JP (2007) Basal ganglia macrocircuits. Prog Brain Res 160:3–7

Terada S, Ishizu H, Yokota O, Tsuchiya K, Nakashima H, Ishihara T, Fujita D, Ueda K, Ikeda K, Kuroda S (2003) Glial involvement in diffuse Lewy body disease. Acta Neuropathol 105:163–169

Thal DR, Del Tredici K, Braak H (2004) Neurodegeneration in normal brain aging and disease. Sci Aging Knowledge Environ 23:1–13

Thiel A, Hilker R, Kessler J, Habedank B, Herholz K, Heiss WD (2003) Activation of basal ganglia loops in idiopathic Parkinson's disease: a PET study. J Neural Transm 110:1289–1301

Tison F (1997) Other causes of parkinsonism. Baillieres Clin Neurol 6:205–218

Tissingh G, Berendse HW, Bergmanns P, De Waard R, Drukarch B, Stoof JC, Wolters EC (2001) Loss of olfaction in de novo and treated Parkinson's disease: possible implication for early diagnosis. Mov Disord 16:41–46

Tofaris GK, Spillantini MG (2005) Alpha-synuclein dysfunction in Lewy body diseases. Mov Disord 20:37–44

Togo T, Iseki E, Marui W, Akiyama H, Ueda K, Kosaka K (2001) Glial involvement in the degeneration process of Lewy body-bearing neurons and the degradation process of Lewy bodies in brains of dementia with Lewy bodies. J Neurol Sci 184:71–75

Tompkins MM, Hill WD (1997) Contribution of somal Lewy bodies to neuronal death. Brain Res 775:24–29

Tracey I (2005) Nociceptive processing in the human brain. Curr Opin Neurobiol 15:478–487

Travagli RA, Hermann GE, Browning KN, Rogers RC (2006) Brainstem circuits regulating gastric function. Annu Rev Physiol 68:279–305

Trojanowski JQ, Lee VMY (1998) Aggregation of neurofilament and α-synuclein proteins in Lewy bodies—implications for the pathogenesis of Parkinson-disease and Lewy body dementia. Arch Neurol 55:151–152

Trojanowski JQ, Lee VMY (2000) "Fatal attractions" of proteins. A comprehensive hypothetical mechanism underlying Alzheimer's disease and other neurodegenerative disorders. Ann NY Acad Sci 924:62–67

Trojanowski JQ, Lee VMY (2002) Parkinson's disease and related synucleinopathies are a class of nervous system amyloidoses. Neurotoxicology 23:457–460

Trojanowski JQ, Lee VMY (2003) Meeting summary—cell biology of Parkinson's disease and related neurodegenerative disorders. Sci Aging Knowledge Environ 13:23

Tu PH, Galvin JE, Baba M, Giasson B, Tomita T, Leight S, Nakajo S, Iwatsubo T, Trojanowski JQ, Lee VM (1998) Glial cytoplasmic inclusions in white matter oligodendrocytes of multiple system atrophy brains contain insoluble alpha-synuclein. Ann Neurol 44:415–422

Uchikado H, Lin WL, DeLucia MW, Dickson DW (2006) Alzheimer disease with amygdala Lewy bodies: a distinct form of alpha-synucleinopathy. J Neuropathol Exp Neurol 65:685–697

Uversky VN, Li J, Fink AL (2001) Evidence for a partially folded intermediate in α-synuclein fibril formation. J Biol Chem 276:10737–10744

Valls-Solé J (2000) Neurophysiological characterization of parkinsonian syndromes. Neurophysiol Clin 30:352–367

Valls-Solé J, Valldeoriola F (2002) Neurophysiological correlate of clinical signs in Parkinson's disease. Clin Neurophysiol 113:792–805

van der Knaap MS, Valk J (1995) Magnetic resonance of myelin, myelination, and myelin disorders, 2nd edn. Springer, Berlin Heidelberg New York

van der Knaap MS, Valk J, Bakker CJ, Schooneveld M, Faber JAJ, Willemse J, Gooskens RH (1991) Myelination as an expression of the functional maturity of the brain. Dev Med Child Neurol 33:849–857

van der Werf, Witter MP, Groenewegen HJ (2002) The intralaminar and midline nuclei of the thalamus. Anatomical and functional evidence for participation in processes of arousal and awareness. Brain Res Brain Res Rev 39:107–140

van Dijk JG, Haan J, Zwinderman K, Kremer B, van Hilten BJ (1993) Autonomic nervous system dysfunction in Parkinson's disease: relationships with age, medication, duration, and severity. J Neurol Neurosurg Psychiatry 56:1090–1095

van Domburg PH, ten Donkelaar HJ (1991) The human substantia nigra and ventral tegmental area. A neuroanatomical study with notes on aging and aging diseases. Adv Anat Embryol Cell Biol 121:1–132

van Hoesen GW, Yeterian EH, Lavizzo-Mourey R (1981) Widespread corticostriate projection from temporal cortex of the rhesus monkey. J Comp Neurol 199:205–219

van Hoesen GW, Morecraft RJ, Vogt BA (1993) Connections of the monkey cingulate cortex. In: Vogt BA, Gabriel M (eds) Neurobiology of cingulate cortex and limbic thalamus. Birkhäuser, Boston, pp 249–284

Veazey RB, Amaral DG, Cowan WM (1982) The morphology and connections of the posterior hypothalamus of the cynomolgus monkey (Macaca fascicularis) II. Efferent connections. J Comp Neurol 207:135–156

Verbaan D, Marinus J, Visser M, van Rooden SM, Stiggelbout AM, van Hilten JJ (2007) Patient-reported autonomic symptoms in Parkinson's disease. Neurology 69:333–341

Vogt BA, Sikes RW (2000) The medial pain system, cingulate cortex, and parallel processing of nociceptive information. Prog Brain Res 122:223–235

Vogt BA, Finch DM, Olson CR (1992) Functional heterogeneity in cingulate cortex: the anterior executive and posterior evaluative regions. In: Jones EG, Peters A (eds) Cerebral cortex. Functional properties of cortical cells, vol 2. Plenum, London, pp 435–443

Vogt BA, Sikes RW, Vogt LJ (1993) Anterior cingulate cortex and the medial pain system. In: Vogt BA, Gabriel M (eds) Neurobiology of cingulate cortex and limbic thalamus. Birkhäuser, Boston, pp 313–344

Volkmann J (2007) Update on surgery for Parkinson's disease. Curr Opin Neurol 20:465–469

Volles MJ, Lansbury PT (2003) Zeroing in on the pathogenic form of alpha-synuclein and its mechanism of neurotoxicity in Parkinson's disease. Biochemistry 42:7871–7878

von Bohlen und Halbach O (2004) Synucleins and their relationship to Parkinson's disease. Cell Tissue Res 318:163–174

Wakabayashi K, Takahashi H (1997a) The intermediolateral nucleus and Clarke's column in Parkinson's disease. Acta Neuropathol 94:287–289

Wakabayashi K, Takahashi H (1997b) Neuropathology of autonomic nervous system in Parkinson's disease. Eur Neurol 38:2–7

Wakabayashi K, Takahashi H, Takeda S, Ohama E, Ikuta F (1988) Parkinson's disease: the presence of Lewy bodies in Auerbach's and Meissner's plexuses. Acta Neuropathol 76:217–221

Wakabayashi K, Takahashi H, Ohama E, Ikuta F (1990) Parkinson's disease: an immunohistochemical study of Lewy body-containing neurons in the enteric nervous system. Acta Neuropathol 79:581–583

Wakabayashi K, Takahashi H, Ohama E, Takeda S, Ikuta F (1991) Lewy bodies in the visceral autonomic nervous system in Parkinson's disease. In: Ikuta F (ed) Neuropathology in brain research. Excerpta Medica, Amsterdam, pp 133–141

Wakabayashi K, Takahashi H, Obata K, Ikuta F (1992) Immunocytochemical localization of synaptic vesicle-specific protein in Lewy body-containing neurons in Parkinson's disease. Neurosci Lett 138:237–240

Wakabayashi K, Takahashi H, Ohama E, Takeda S, Ikuta F (1993) Lewy bodies in the visceral autonomic nervous system in Parkinson's disease. Adv Neurol 60:609–612

Wakabayashi K, Hansen LA, Masliah E (1995) Cortical Lewy body-containing neurons are pyramidal cells. Laser confocal imaging of double-immunolabeled sections with anti-ubiquitin and SMI32. Acta Neuropathol 89:404–408

Wakabayashi K, Hayashi S, Kakita A, Yamada M, Toyoshima Y, Yoshimoto M, Takahashi H (1998) Accumulation of α-synuclein/NACP is a cytopathological feature common to Lewy body disease and multiple system atrophy. Acta Neuropathol 96:445–452

Wakabayashi K, Engelender S, Yoshimoto M, Tsuji S, Ross CA, Takahashi H (2000a) Synphilin-1 is present in Lewy bodies in Parkinson's disease. Ann Neurol 47:521–523

Wakabayashi K, Hayashi S, Yoshimoto M, Kudo H, Takahashi H (2000b) NACP/α-synuclein-positive filamentous inclusions in astrocytes and oligodendrocytes of Parkinson's disease brains. Acta Neuropathol 99:14–20

Wakabayashi K, Tanji K, Mori F, Takahashi H (2007) The Lewy body in Parkinson's disease: molecules implicated in the formation and degradation of alpha-synuclein aggregates. Neuropathology 27:494–506

Walker LC, LeVine H (2001) The cerebral proteopathies. Neurodegenerative disorders of protein conformation and assembly. Mol Neurobiol 21:83–95

Waseem S, Gwinn-Hardy K (2001) Pain in Parkinson's disease. Postgrad Med 110:1–5

Wenning GK, Colosimo C, Geser F, Poewe W (2004) Multiple system atrophy. Lancet Neurol 2:93–103

Westermann B, Wattendorf E, Schwerdtfeger U, Husner A, Fuhr P, Gratzl O, Hummel T, Bilecen D, Welge-Lüssen A (2008) Functional imaging of the cerebral olfactory system in patients with Parkinson's disease. J Neurol Neurosurg Psychiatry 79:19–24

Wichmann T, DeLong MR (1993) Pathophysiology of Parkinsonian motor abnormalities. Adv Neurol 60:53–61

Wichmann T, DeLong MR (1997) Functional and pathophysiological models of the basal ganglia. Curr Opin Neurobiol 6:751–758

Wichmann T, DeLong MR (1998) Models of basal ganglia function and pathophysiology of movement disorders. Neurosurg Clin N Am 9:223–236

Wichmann T, DeLong MR (2003) Functional neuroanatomy of the basal ganglia in Parkinson's disease. Adv Neurol 91:9–18

Wichmann T, DeLong MR (2006) Deep brain stimulation for neurologic and neuropsychiatric disorders. Neuron 52:197–204

Willis WD, Westlund KN (1997) Neuroanatomy of the pain system and of the pathways that modulate pain. J Clin Neurophysiol 14:2–31

Winge K, Fowler CJ (2006) Bladder dysfunction in parkinsonism: mechanisms, prevalence, symptoms, and management. Mov Disord 21:737–745

Winn P, Brown VJ, Inglis WL (1997) On the relationships between the striatum and the pedunculopontine tegmental nucleus. Crit Rev Neurobiol 11:241–261

Wirch JL, Wolfe LA, Weissgerber TL, Davies GA (2006) Cold pressor test protocol to evaluate cardiac autonomic function. Appl Physiol Nutr Metab 31:235–243

Witter MP, Room P, Groenewegen HJ, Lohman AHM (1988) Reciprocal connections of the insular and piriform claustrum with limbic cortex: an anatomical study in the cat. Neuroscience 24:519–539

Wolters EC, Braak H (2006) Parkinson's disease: premotor clinico-pathological correlations. J Neural Transm 70:309–319

Wolters EC, Francot CMJE (1999) The concept of mental dysfunction in Parkinson's disease. In: Wolters EC, Scheltens P, Berendse HW (eds) Mental dysfunction in Parkinson's disease. Netherl Acad Pharm Prod, Utrecht, pp 35–48

Wolters EC, Francot C, Bergmans P, Winogrodzka A, Booij J, Berendse HW, Stoof JC (2000) Preclinical (premotor) Parkinson's disease. J Neurol 247:103–109

Wurtz RH, Hikosaka O (1986) Role of the basal ganglia in the initiation of saccadic eye movements. Prog Brain Res 64:175–190

Yelnik J, Percheron G (1979) Subthalamic neurons in primates: a quantitative and comparative analysis. Neuroscience 4:1717-1743

Yelnik J, Francois C, Percheron G, Heyner S (1987) Golgi study of the primate substantia nigra. I. Quantitative morphology and typology of nigral neurons. J Comp Neurol 265:455–472

Yelnik J, Francois C, Percheron G, Tandé D (1991) Morphological taxonomy of the neurons of the primate striatum. J Comp Neurol 313:273–294

Yeterian EH, Pandya DN (1991) Prefrontostriatal connections in relation to cortical architectonic organization in rhesus monkeys. J Comp Neurol 312:43–67

Yeterian EH, Pandya DN (1994) Laminar origin of striatal and thalamic projections of the prefrontal cortex in rhesus monkeys. Exp Brain Res 99:383–398

Yu S, Li X, Liu G, Han J, Zhang C, Li Y, Xu S, Liu C, Gao Y, Yang H, Uéda K, Chan P (2007) Extensive nuclear localization of alpha-synuclein in normal rat brain neurons revealed by a novel monoclonal antibody. Neuroscience 145:539–555

Yung KKL, Smith AD, Levey AI, Bolam JP (1996) Synaptic connections between spiny neurons of the direct and indirect pathways in the neostriatum of the rat: evidence from dopamine receptor and neuropeptide immunostaining. Eur J Neurosci 8:861–869

Zahm DS (2000) An integrative neuroanatomical perspective on some subcortical substrates of adaptive responding with emphasis on the nucleus accumbens. Neurosci Biobehav Rev 24:85–105

Zaja-Milatovic S, Milatovic D, Schantz AM, Zhang J, Montine KS, Samii A, Deutch AY, Montine TJ (2005) Dendritic degeneration in neostriatal medium spiny neurons in Parkinson's disease. Neurology 64:545–547

Zarow C, Lyness SA, Mortimer JA, Chui HC (2003) Neuronal loss is greater in the locus coeruleus than nucleus basalis and substantia nigra in Alzheimer and Parkinson diseases. Arch Neurol 60:337-341

Zecca L, Zucca FA, Albertini A, Rizzio E, Fariello RG (2006) A proposed dual role of neuromelanin in the pathogenesis of Parkinson's disease. Neurology 67:8–11

Zilles K (2004) Architecture of the human cortex. In: Paxinos G, Mai JK (eds) The human nervous system, 2nd edn. Elsevier, London, pp 997–1060

Zola-Morgan S, Squire LR (1993) Neuroanatomy of memory. Annu Rev Neurosci 16:547–563

Zweig RM, Jankel WR, Hedreen JC, Mayeux R, Price DL (1989) The pedunculopontine nucleus in Parkinson's disease. Ann Neurol 26:41–46

Zweig RM, Cardillo JE, Cohen M, Giere S, Hedreen JC (1993) The locus ceruleus and dementia in Parkinson's disease. Neurology 43:986–991

Made in United States
Orlando, FL
15 July 2026

84345142R00079